Handbook of Online and Near-real-time Methods in Microbiology

Handbook of Online and Near-real-time Methods in Microbiology

Editors

Maximilian Lackner

Institute of Advanced Engineering Technologies
University of Applied Sciences FH Technikum Wien
Vienna, Austria

Philipp Stadler

TU Wien, Centre for Water Resource Systems
Vienna, Austria

Wilhelm Grabow

Formerly Head, Department of Medical Virology,
University of Pretoria, Pretoria, Republic of South Africa

CRC Press
Taylor & Francis Group
Boca Raton London New York

CRC Press is an imprint of the
Taylor & Francis Group, an **informa** business
A SCIENCE PUBLISHERS BOOK

Cover photograph reproduced by kind courtesy of DI Wolfgang Vogl, VWMs GmbH, 2017.

CRC Press
Taylor & Francis Group
6000 Broken Sound Parkway NW, Suite 300
Boca Raton, FL 33487-2742

Printed on acid-free paper
Version Date: 20170508

International Standard Book Number-13: 978-1-4987-6402-5 (Hardback)

Library of Congress Cataloging-in-Publication Data

Names: Lackner, Maximilian, editor. | Stadler, Philipp, 1975- editor. |
Grabow, W. O. K., editor.
Title: Handbook of online and near-real-time methods in microbiology /
editors, Maximilian Lackner, Institute of Advanced Engineering
Technologies, University of Applied Sciences FH Technikum Wien, Vienna,
Austria, Philipp Stadler, TU Wien, Centre for Water Resource Systems,
Vienna, Austria, Wilhelm Grabow, Formerly Head, Department of Medical
Virology, University of Pretoria, Pretoria, South Africa.
Description: Boca Raton, FL : Taylor & Francis Group, 2017.
| Includes bibliographical references and index.
Identifiers: LCCN 2017018989| ISBN 9781498764025 (hardback : alk. paper) |
ISBN 9781498764049 (e-book : alk. paper)
Subjects: LCSH: Microbiology--Handbooks, manuals, etc. |
Microbiology--Methodology. | Microorganisms--Detection.
Classification: LCC QR72.5 .H36 2017 | DDC 579--dc23
LC record available at https://lccn.loc.gov/2017018989

Visit the Taylor & Francis Web site at
http://www.taylorandfrancis.com

and the CRC Press Web site at
http://www.crcpress.com

Preface

Water, as well as other fluids intended for human use, are characterized by many parameters, most of which can be measured directly with high temporal resolution, high accuracy and low effort. Take temperature, conductivity, turbidity, density and composition—there is a large number of techniques available to probe small and large samples to obtain a value in near-real-time and/or to rapidly control a process.

One crucial parameter is the microbiological quality of utilized liquids. The presence of microorganisms in a sample can still not be determined readily. Today researchers and practitioners in the fields of hygiene, healthcare and other disciplines still rely on culture methods established more than a century ago to assess the microbiological quality of water. These culture-base microbial assays, that require the plating of a sample on growth agar, incubating and counting colonies after typically 24 hours, are still state-of-the-art. This process demands elaborate sampling and laboratory work, is time- and staff consuming and consequentially costly. It takes more than a working day to gain results that are critical for a health-related surveillance of water resources of liquid systems. Therefore, these microbiological standard assays can barely be integrated into early-warning systems and dynamic process control.

Online and near-real-time measurements of traditional physical parameters have added great value to many fields of environmental research as well as various industries. They can be seen as the core of the industry where, a new trend in manufacturing is poised at increasing flexibility and reducing costs. In environmental research the on-site and near-real-time monitoring of physico-chemical parameters of water is well established, commonly used and a core component of sustainable water resource management. For microbial and biochemical parameters, however, there is still a lack of technologies and assays that allow such a high temporal resolution of measurement. Decisive technological progress occurred in the last few years, generating new methods and assays that are able to highlight the microbial quality of water and liquids in near-real-time. While some of these methods and related research are still emerging, they have enormous potential to enhance health-related assessment of environmental, medical or industrial systems. It can be expected that near-real-time microbial methods will be the cornerstone of water quality monitoring and industrial process control in the near future. Imagine, for instance, a drinking water or sewage treatment plant, where disinfection is carried out. Today, chemicals need to be added in excess to be 'on the safe side', or UV light has to be provided at installed capacity. Now if an operator were to instantly measure the microbiological contamination level of either inlet or outlet stream, he could automatically adjust the degree of required disinfection to reach the target level, thereby saving cost, relieving the environment and adding safety to the process, now closely monitored. This is just one of the many potential use cases of rapid microbiological sensors.

In this handbook, leading experts contribute review articles on frontiers in microbiological detection research—emerging and state-of-the-art online and near-real-time methods of microorganism detection and indication.

This book presents cutting edge research, developments and applications with regard to online and near-real-time microbiology, where 'near-real-time' is understood as obtaining a quantitative or at least qualitative measurement result within 15–30 minutes, which is really fast as compared to traditional methods that show a time lag often in excess of 24 hours.

The aim was to obtain a balance between chapters from industry and contributions from academia, covering the broad field of microbiological quality of water and liquids in natural, industrial and medical systems.

In their chapter titled 'Rapid, Automated and Online Detection of Indicator Bacteria in Water', Trude Movig and others describe how microbiological water quality can be monitored by analyzing fecal indicator bacteria, like thermotolerant coliforms and *E. coli* based on their enzymatic activities; chromogenic substances are utilized for the detection of β-D-galactosidase and β-D-glucuronidase (GLUC) enzyme activities, respectively, to indicate the presence of these bacteria, which are well-established indicators for fecal contamination that can carry other pathogenic microorganisms. The chapter discusses analyzers that are in market today and compares their performance based on results generated in 0.25 to 2 hours as compared to automated growth methods that have a typical analysis time of 12–17 hours.

The next chapter titled 'Real-time Monitoring of Enzymatic Activity in Water Resources' by Philipp Stadler and others presents the application of a novel concept for automated online monitoring of enzymatic activity in different water resources, specifically alluvial porous groundwater, karstic groundwater and surface water. Two commercially available prototypes for beta-D-glucuronidase (GLUC) activity determination were subjected to an extensive field trial in an Austrian catchment area. The experimental results of Stadler et al. show that on-site measured GLUC signals reflect the characteristic transport and discharge dynamics of the observed catchment. Reference analytics by means of culture-based *E. coli* indicate that GLUC measurements are not a proxy for standard microbiological assays and that relations of GLUC activity and culture-based *E. coli* depend on the observed habitat. Comparison of GLUC measurements gathered with independently-constructed prototypes showed comparable results.

'Advances in Electrochemically Active Bacteria Physiology and Ecology' by Ana C. Marques and others presents electrochemically active bacteria (EAB), which constitute a very interesting and promising class of bacteria. However, today's screening methods are based on MFC (microbial fuel cell) engineering principles, which are relatively slow (~5 to 6 days) and expensive. So, the development of rapid and simple screening methods using low cost and available materials are today a key issue to aid in the better understanding of such types of bacteria, thus allowing further refining in the performance of electrochemically active bacteria for use in biotechnological applications.

Sevcan Aydin presents the 'Application of Quantitative Real-time PCR for Microbial Community Analysis in Environmental Research' in his chapter. Polymerase chain reaction (PCR) operates by extracting nucleic acids and their enzymatic amplification of certain genes from the complex genomic DNA of environmental samples. Aydin discusses the molecular approach of PCR for near-real-time bacterial detection.

The next chapter addresses the measurement of bacteria in a completely different fluid than water—Gregor Tegl and others describe the '*In Vitro* Diagnostics for Early Detection of Bacterial Wound Infection', where they use enzymes produced by the infected wound to rapidly infer bacterial colonization through a color change with an *in vitro* diagnostic (IVD) tool. Results are obtained within a compelling 15 minutes. Bacterial contamination is the basis of wound infection and affects about 10 per cent of all postoperative wounds. A high risk of infection is also reported for chronic wounds.

Jacobo Paredes and others have contributed the chapter 'Biofilm Impedance Monitoring'. Biofilms are a challenge to control and their early detection is an advisable strategy, for which the authors demonstrate impedance microbiology as a suitable approach.

Maximilian Lackner has written the chapter 'Rapid Microbial Water Quality Measurement by Automated Determination of the Fecal Indicator Bacterium *Escherichia coli*', which reviews the measurement principles, opportunities and limitations of automated microbiological measurements for process control.

Samendra P. Sherchan presents the final chapter on real-time monitoring of microorganisms in potable water using online sensors for use in SCADA (supervisory control and data acquisition) systems in large-scale water quality monitoring programs. Multi-angle light scattering (MALS), intrinsic fluorescence and enzyme-based technology are compared and discussed, also with reference to traditional sensors and influencing factors.

The purpose of this Handbook is to serve its readers as a handy reference and give momentum to their thoughts on novel applications of the present emerging online and near-real-time methods in microbiology. It also contains a comprehensive glossary in which the most relevant terms of this novel field are defined. The new and up-to-now fairly narrow field of rapid and automated microbial assays is likely to gain significance and cost an impact on a global scale in the near future. This Handbook is the first of its kind, giving an up-to-date insight into modern monitoring and measuring techniques.

The editors and authors wish their readers many interesting insights after reading this Handbook.

Vienna, January 21, 2017

Maximilian Lackner
Philipp Stadler
Wilhelm Grabow

Contents

1

Rapid, Automated and Online Detection of Indicator Bacteria in Water

*Trude Movig, Henrik Braathen** and *Helene Stenersen*

1. Introduction

Having access to safe drinking water is a basic human right and essential to our health (UN 2010). In both developing and developed countries, microbial hazards are a prime source of concern when assessing drinking water. Furthermore, recreational water of poor quality may also pose a risk to human health. Outbreaks of water-borne diseases occur frequently and some of them remain undetected, constituting a background level of disease in human populations. Pathogenic microorganisms are commonly distributed in water via the feces of humans and occasionally of some animals. Their verification is commonly based on the analysis of fecal indicator organisms, such as *E. coli* and/or thermotolerant coliforms, the presence of which indicates a recent fecal pollution (WHO 2011, EU 2006). Thermotolerant coliforms are often referred to as fecal coliforms (FC). To detect and respond to a contamination episode, it is crucial to have the shortest possible amount of time beginning from sampling to result. Ideally, the process is fully automated with a continuous analysis of the water and results are presented in real-time. Only the target organism is detected at

Colifast AS, Strandveien 35, P.O.Box 31, 1324 Lysaker, Norway. E-mail: post@colifast.no
* Corresponding author: hb@colifast.no

the smallest amount possible. By use of fully automated methods, there is no need for manual water collection, transportation of samples, preparation and analysis in a laboratory. Furthermore, the analysis can be performed outside of work hours. Automated sampling at the site further reduces the total time to result, as analysis will start directly after sampling. It also reduces the risk of possible degradation of sample due to the many steps involved in manual sampling and analysis (Fig. 1).

Fully automated water analysis enables increased test frequency and generates additional quality data to present a better overview of variations and pollution episodes. This would be beneficial for many waterworks as the traditional manual sampling and lab analysis results are limited to a few samples a month (EU 1998). The following overview of available analyzers only focuses on the fully automated units. There are also available systems for automated lab analysis with manual addition of sample but this chapter only describes the analyzers which are able to automatically grab the sample and perform analysis without assistance from an operator.

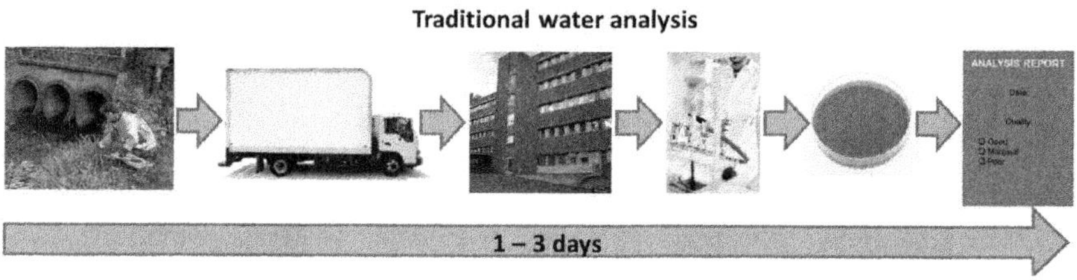

Figure 1. Traditional sampling and analysis including manual sampling, transportation, sample preparation, analysis and interpretation of results and reporting.

2. Technology Background

2.1 Indicators

In general, the quality of drinking water is determined by analyses of physical, chemical and microbiological parameters. A common source of water-borne disease is consumption of drinking water contaminated with human or sometimes animal excreta (Cabral 2010, CDC 2015). A large number of different pathogens might be present in feces, such as viruses, bacteria and parasites, often in low concentrations. Individual monitoring of them is not possible for waterworks due to high costs, complicated sampling procedures and laboratory analysis. Instead, indicator organisms, such as thermotolerant coliforms and *E. coli*, are monitored on a routine basis (APHA 2005). These organisms are usually non-pathogenic and present in high numbers in the feces of humans and warm-blooded animals. Their presence in water indicates a recent fecal contamination and the possible presence of enteric pathogens. *E. coli* is considered the best indicator, whereas total coliforms are an acceptable, but less reliable indicator as some of these bacteria can live in the soil. Thermotolerant coliforms are considered as a more reliable indicator compared to the total coliforms but is less specific when compared to *E. coli*. The larger group of bacteria detected by total coliform analysis may be a more sensitive and good indicator of contamination of drinking water due to leakage in the distribution system. The detection of *E. coli*, or alternatively thermotolerant coliforms, is typically done in 100 ml samples and the accepted value of drinking water is 0/100 ml (WHO Guidelines). The fecal indicators are also widely used to indicate the health risks for recreational water as high levels of fecal bacteria are related to illness for people in contact with the water. The accepted level of indicators is higher for this water as it is not used for consumption. Guidelines for recreational waters typically set the values to 200–1000/ 100 ml (EU Guidelines). For decades, the correlation between these indicators and harmful fecal contamination has been debated. However, they continue to be a central part of the assessment of hygienic water quality in national standards and international guidelines worldwide.

2.2 History of growth media and instruments

Traditionally, the detection of coliforms and *E. coli* was based on their ability to ferment lactose with the production of gas and acid by using the indole reaction for visual color change to confirm the results. Later, the colored product of the o-nitrophenyl-β-d-galactopyranoside (ONPG) reaction was introduced for detecting fecal indicators. The bacteria were grown on the surface of a solid nutrient-rich agar, usually containing yeast extract and peptone. These methods were both time and labor intensive, often including membrane filtration and confirmation steps and the results were only available after 24–72 hours (WHO 2001). In the 1990s, rapid and simplified techniques for the monitoring of water quality (Berg and Fiksdal 1988, Edberg et al. 1988) were introduced. These new techniques used a liquid growth medium with direct addition of the sample to the medium. The aim was to decrease the time to detection of the target organisms and to improve the sensitivity of the assay, so that the time for the result was further reduced. Also, it was stated that the methods should be easier to perform in order to lessen both labor and costs. The focus was directed in two different directions:

- The development of methods
- The development of instrumentation

And for some, the attention was on both. There was also a focus on development of an enzymatic rapid method that could be used for rapid screening (Fiksdal et al. 1994, Eckner et al. 1999, George et al. 2000). This resulted in chromogenic substances that utilize the detection of β-D-galactosidase and β-D-glucuronidase enzyme activities to indicate the presence of total coliforms or *E. coli*, respectively. These are now widely used in public health microbiology (Edberg et al. 1990).

The methods are easy to use and are in general based on the detection of a visual end-product. The analysis time for drinking water samples is still between 14–24 hours (Tryland et al. 2015). Reduction in analysis time may be achieved by use of sensitive instrumentation to detect released chemiluminescence (Van Poucke and Nelis 1995), colored (Apte et al. 1995) or fluorescent end-products (Tryland et al. 2001). These are incorporated into substrates usually based on galactosides, including lactose (Fig. 2). The substrates with a chemiluminescent end-product are typically 3-(4-methoxyspiro)-1,2-dioxetane-3,2′-tricyclo[3.3.1.1]decan-(4-yl)phenyl-β-d-galactopyranoside (AMPGD). The substrates with a colored end-product are usually 2-nitrophenyl β-D-galactopyranoside (OPNG) and

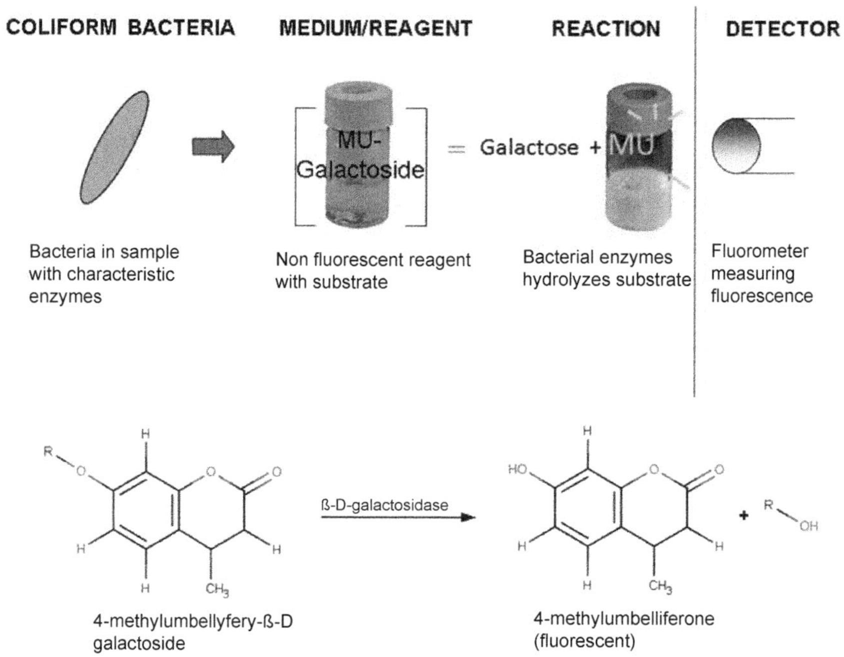

Figure 2. The reaction by coliform characteristic enzymes which hydrolyze specific substrates, leading to production of a fluorescent end-product. The end-product can be measured by a detector or detected visually by use of an UV lamp.

substrates with fluorescent end-product are usually 4-methylumbelliferone β-D-galactopyranoside (MU-Gal). The detectable end-product is released after reaction by the bacterial enzymes which cleave the glycosidic bond between galactose and the end-product. Similar glucuronide-based substrates are also available for specific *E. coli* glucuronidase activity.

Colifast in Norway is the first company known to develop an automated instrument for monitoring indicator bacteria in water by use of these substrates. The company based the automated analysis on in-house developed growth media containing specific fluorogenic substrates (Berg 1991, 1993). In the mid 90s, the Colifast analyzer 100 was introduced in the market and some years later it was followed by Colifast CA (Fig. 3). Both the systems were semi-automated and the operator added sample and media to test vials prior to the automated analysis.

In the year 2000, the first Colifast CALM was introduced. This is a fully-automated system which is connected directly to a sample line at a designated site (Demowatercoli 2005). The CALM system generates quantitative results and can run both a 2-hour rapid screening test and two different growth methods (6–12 hours). It can run up to 76 samples per run. Another automated system was also introduced in the marked in this period. The French company, Seres, developed a system named Colilert 3000. This

Figure 3. The first generation of semi-automated analyzers introduced in the 1990s, the Colifast CA-100. The CA-100 was a laboratory instrument and could automatically analyze 96 samples after sample preparation.

system is primarily used for indicator presence/absence testing of 100 ml drinking water samples (Zuckerman et al. 2005) with an analysis time of a maximum of 17 hours. It can analyze up to 35 samples per run. Other systems have been launched, but most of these have been discontinued due to quality problems or economical problems for the manufacturer. By 2015, several new automated and semi-automated systems were introduced in the market. There would probably also be similar systems under development or not internationally available, but these are not presented. The fully automated systems available today are the BACTcontrol from micro LAN, the Coliminder by Vienna Water Monitoring, the Coli 3000 by Seres and the ALARM and CALM by Colifast.

2.3 Automated formats

When the bacteria are transferred to a new environment, e.g., a liquid medium, growth will normally occur after a time-period, called the lag-phase. During the lag-phase, the bacteria will adjust their metabolism to the new environment/medium and sub-lethally damaged cells may be resuscitated. Subsequently, the bacterial population undergoes exponential growth. The fluorescence/color development of samples analyzed by an instrument in general follows a pattern similar to the bacterial growth, with the lag-phase followed by the exponential phase. If the samples contain high levels of coliform bacteria, a linear increase in fluorescence/color is in general seen in the lag-phase (Fig. 4).

2.3.1 PA (presence/absence) format[1]

If there are a low number of bacteria in the water samples, it is normal to examine 100 ml of water and investigate if there is presence or absence of coliform bacteria in these samples. The sample is mixed with the medium and after a defined incubation time, adequate for bacterial growth, the color/ fluorescence of the liquid mix is measured. Strong color/fluorescence above a selected threshold value indicates a presence result and no or low color/ fluorescence indicates an absence result. The direct addition of the sample to the medium without a filtration step may occasionally lead to inhibition of bacterial growth if the water volume contains inhibitory elements.

[1] The presented analysis formats, benefits and limitations are based on experience from Colifast AS. The company has used these different formats in numerous applications during the past 20 years.

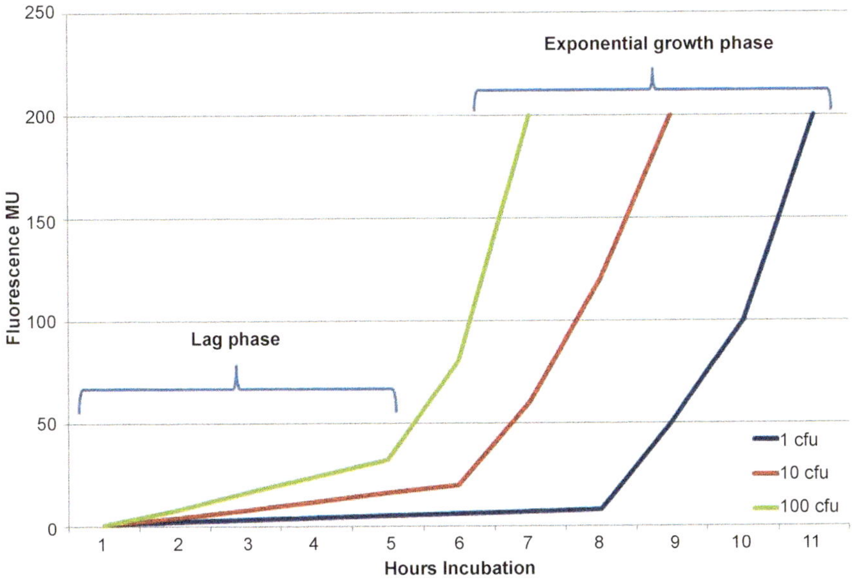

Figure 4. Fluorescence from river water samples. The fluorescence follows a pattern similar to the bacterial growth and consists of a lag-phase and an exponential phase. High numbers of coliform bacteria give an early increase in the fluorescence; lower numbers of coliform bacteria require a longer incubation time before a logarithmic increase in fluorescence is seen.

2.3.2 MPN (most probable number) format[1]

The MPN format is a method to quantify bacteria growing in the medium and the format is commonly performed by laboratories using the ISO 9308-2:2012 standard method. The MPN format is based on the PA format, but has several duplicates. The water sample is distributed into several vials/tubes containing the growth medium, for example 5 x 10 ml. By inserting the sample volume and counting the number of tubes that gave a positive reaction on bacteria, the MPN-value can be found to use a statistical calculated table. These tables are usually based on the Thomas simple formula. For the common fecal indicators, the MPN-value shows the number of bacteria per 100 ml sample.

Thomas' formula:

$$MPN/100\ ml = \frac{no.\ of\ \ positive\ vials \quad \times \quad 100}{\sqrt{(ml\ sample\ in\ negative\ vials \quad \times \quad ml\ sample\ in\ all\ vials)}}$$

[1] The presented analysis formats, benefits and limitations are based on experience from Colifast AS. The company has used these different formats in numerous applications during the past 20 years.

2.3.3 TTD (time to detect) format[1]

A high number of target bacteria in the sample will lead to a quicker increase in fluorescence compared to a sample with a low number of bacteria. By frequent measurement of fluorescence/color in the sample mixed with a medium, this can be used to determine the level of bacteria in the sample by the number of hours to get a positive fluorescence/color signal after an exponential (growth) increase in fluorescence, above a selected threshold value. This time is called time to detect (TTD) and the bacterial number is estimated, using a semi-quantification table. The semi-quantification table is based on empirical data. This method is known to be quite rough and would normally need local calibration for method adjustment. Variation in the stress level of the bacteria, shifting distribution of bacteria strains and other local environmental factors are known to alter the time to detect.

2.3.4 Enzyme activity format[1]

Enzyme activity during the first minutes and hours after addition of sample to the liquid medium can be measured. This is often referred to as the MU-production (MUP) method and is based on linear fluorescence development. MU is the described fluorescent end-product 4-methylumbelliferone. The results are calculated as the production of MU per hour (ppb MU/h), reflecting the β-galactosidase or β-glucuronidase activity of the water samples prior to any growth. The ppb MU/h result can be converted to a corresponding coliform number using a local correlation curve. The MUP method can be used for rapid screening of sewage and surface water containing medium to high levels of fecal contamination (e.g., contaminated rivers). As this method does not wait for bacterial growth, it will not discriminate between viable and non-viable bacteria. Extracellular enzymes and enzymes from non-culturable bacteria can therefore interfere with the results.

3. Analyzers in the Market Today

Many of today's waterworks wish for a more extensive control and testing of water with an increased frequency of measurements and quicker results. Their water cannot be held in quarantine, such as with food products, and is normally consumed before the testing is completed. So it is crucial to have

[1] The presented analysis formats, benefits and limitations are based on experience from Colifast AS. The company has used these different formats in numerous applications during the past 20 years.

data about the quality of the water as frequently and rapidly as possible. *E. coli* typically survive in water for a few days. So measuring levels of *E. coli* under standard conditions, e.g., once a week, is not likely to reveal all the contamination episodes. In order to perform testing during day and night all days of the week, the testing would have to be automated. Such a method must be rapid, automated and online. The ideal situation would be a continuous real-time measurement directly on the process line. Furthermore, the process of water treatment could be adjusted in a system with a tight feedback loop from the rapid online bacterial monitoring, in addition to other central process parameters. The results must be sensitive with no false negatives or positives and specific, i.e., detect only target organisms. The system should be fully automated from sampling to result and easy to use by the staff with no laboratory training. These methods are typically more expensive in terms of instrument purchase and consumables than the traditional methods. However, in all cases, the automated methods require less manpower and their cost will therefore be reduced, including costs of sampling in remote locations (Fig. 5).

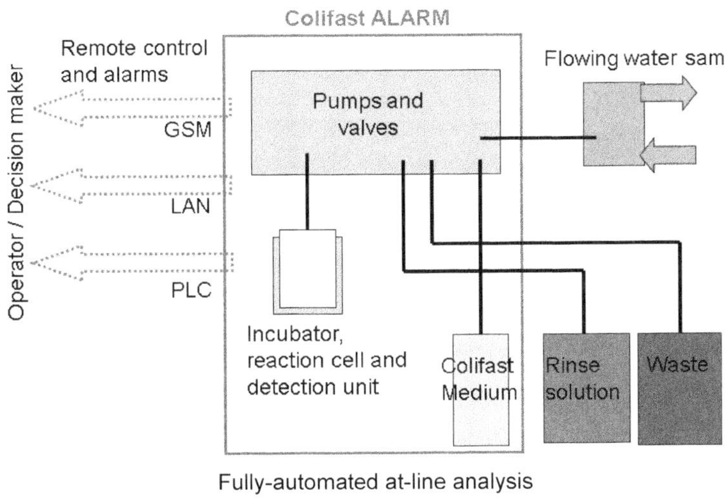

Figure 5. A schematic overview of the Colifast ALARM system showing water sampling, main components and interface options for transmitting results to the operator: Global System for Mobile Communication (GSM), Local Area Network (LAN) and Programmable Logic Controller (PLC).

3.1 Automated growth methods

All the accepted standard laboratory methods used today for coliform detection in water are based on bacterial growth in combination with specific characteristics of the target bacteria. The automated growth methods utilize the same principles for detection.

3.1.1 Colifast CALM

Colifast AS has developed a technology for analyzing bacteria in different kinds of water, such as raw water, fresh water, seawater, drinking water, bottled water and process water. The main components of CALM (Colifast At-Line Monitor) are two incubator blocks (80 positions/wells), a random-access sampler XYZ-robot that automates liquid handling, an analyzer module with valves and pumps and a detector composed of a fluorometer, a light source and a flowcell. The system is operated by an integrated computer with CALM software to automatically grab the sample, add it to wells containing the medium by the robot and incubate the sample. The system then measures sub-samples from the wells at pre-programmed intervals. The software calculates the number of bacteria and reports results by an industrial interface, by a network connection and/or SMS, as soon as the results are available. The technology can detect *E. coli*, total coliforms and thermotolerant coliforms in water samples, plus *Pseudomonas aeruginosa*. The Colifast total coliform and thermotolerant coliform growth media contain the substrate 4-methylumbelliferyl (MU)-β-D-galactoside, and this substrate is hydrolyzed by the enzyme β-galactosidase that is present in coliform bacteria. The Colifast *E. coli* medium contains the substrate 4-Methylumbelliferyl (MU)-β-D-glucuronide, which is hydrolyzed by the enzyme β-D-glucuronidase that is present in *E. coli*. The fluorescent product MU is produced as a result of the hydrolysis reaction. The Colifast *Pseudomonas aeruginosa* medium contains substrate L-arginine-7-amido-4-methylcoumarin (L-arg-AMC), which is hydrolyzed by an amino peptidase enzyme present in *Pseudomonas aeruginosa*. The fluorescing product 7-amino-4-methylcoumarin (AMC) is produced as a result of the hydrolysis reaction. The CALM method combines the Colifast selective growth media with an automated analyzer. The different bacteria types are separated by specific enzyme activity and by different incubation temperatures in the analyzer. Total coliforms, *E. coli* and *Pseudomonas aeruginosa* are incubated at 37°C and thermotolerant coliforms are incubated at 44°C. Inhibitors in the medium prevent growth of non-target bacteria. It takes 2–12 hours to get the analysis results from the Colifast method, depending on the amount of bacteria present and the format of the analysis (Fig. 6). The CALM system can run all the four formats (PA, MPN, TTD and MUP) and would usually be customised for the selected application (Braathen et al. 2005).

Figure 6. The fully automated CALM system as installed in a river-monitoring station in 2006. The system automatically grabs a sample and starts analysis at pre-programmed intervals. Sample collector flask with continuous water-flow is seen on the right side.

3.1.2 Seres/Colilert 3000

The Seres Coli 3000 instrument comes in two editions—one automatic and the other semi-automatic. This review will only discuss the automatic model. The system consists of pumps and valves, detector units and control/interfacing units. This system has three separate tempered reaction chambers with separate addition of sample and medium and a detection unit on each chamber. After analysis, the chambers are automatically rinsed and reused. This system uses the IDEXX Colilert medium based on the patented Defined Substrate Technology for analysis with a maximum of 17 hours for detection of one indicator bacterium/100 ml sample.

The instrument can take out water samples with a minimum of 6-hour interval. The complete measuring cycle takes 18 hours. The culture medium used in the Seres Coli 3000 instrument is also based on specific enzyme activity and the hydrolyzes and release of end-products that color the medium yellow and/or blue (fluorescence). This coloration is detected by the Seres Coli 3000 analyzer. The instrument can either detect total coliforms, *E. coli* or thermotolerant coliforms in sea or fresh water down to 1 cfu/100 ml using a PA format with an early warning option, if the sample contains elevated bacterial levels (Fig. 7). The analysis is controlled by an onboard data and control unit and results can be reported by industrial interface (Seres Environment, France).

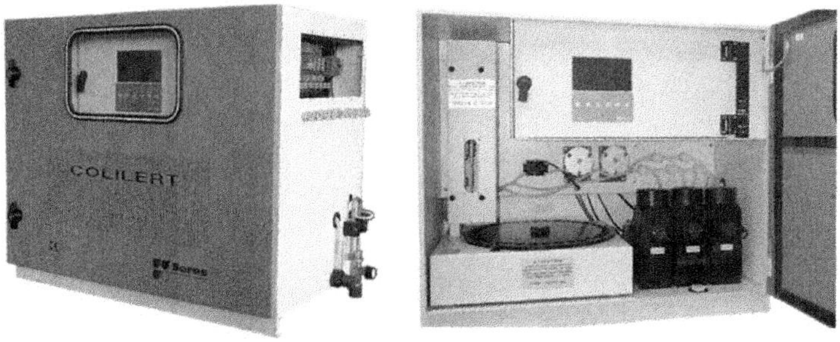

Figure 7. The Seres Coli 3000 including the technical units for medium and sample handling and the 3 reaction/analysis chambers.

3.1.3 Colifast ALARM

The Colifast ALARM (At-Line Automated Remote Monitor) is an automated system for detection of total coliforms, thermotolerant coliforms or *E. coli* in 100 ml water samples. The analysis and detection is based on the Colifast selective growth medium, including the described fluorogenic substrates. Colifast ALARM consists of pumps and valves for liquid handling, one incubator/reaction cell connected to a detector, control/interfacing units and an integrated touchscreen computer. The Colifast ALARM automatically collects the water sample at programmed intervals with a maximum sampling rate of 15 hours. The system can automatically send results to the control room/laboratory via LAN remote control, by industrial interface or by mobile phone network (SMS). In addition, the Colifast ALARM measures the turbidity (cloudiness) level in water samples (Fig. 8).

This system can detect down to one target bacterium per 100 ml and the results are obtained within 14 hours (*E. coli*) with early warning of high bacterial levels at 6 hours onwards (Colifast AS, Norway).

3.2 Automated non-growth methods

The culture methods commonly used by laboratories needs 18–24 hours to be completed, thereby only recording a past event. There are no results that could be used for decision making and pro-active protection of public health (Tryland et al. 2002). The emergence of automated and rapid methods has

Figure 8. The Colifast ALARM system as installed at a water treatment plant for monitoring efficiency of filtration process.

reduced the time to 12 hours (Colifast CALM) and further development of methods that aims real-time results continues.

The major disadvantage for real-time analyses is that they fail to discriminate between non-viable and viable bacteria. The presence of dead bacteria can lead to false positive results that may have severe consequences for the user. Further development or use of excising technologies that only detect live bacteria will be the solution to this problem.

3.2.1 Enzymatic

The enumeration of indicator bacteria in water by the use of specific enzyme activities without bacterial growth has been available for years. Furthermore, they have been incorporated in automated analyzers since the early 2000s, when trials were performed by Colifast AS to develop a semi-automated early warning system for monitoring of fecal contamination (Tryland et al. 2002). Recently, other methods and instruments have emerged (Ryzinska-Paier 2014). These systems can automatically grab a sample, perform a rapid analysis in less than 3 hours, wash out the system and start analyzing a new sample.

These technologies consist of a sensitive analysing instrument and a reagent containing the specific fluorogenic substrates. By measuring the changes in fluorescence levels during the short lag phase (before bacterial growth), the time taken for these methods is significantly reduced. Viable but not-culturable (VBNC) cells can have an enzyme activity that allows for detection in a rapid enzymatic assay, but not by the traditional growth methods. These cells might suffer from sub-lethal damages due to disinfection by agents, such as chlorine or UV-radiation. Furthermore, dead cells might leak enzymes that can be involved in the production of fluorescent end-products that will be detected by the system as a signal of present coliforms. However, the traditional reference methods are based on the cultivation of these bacteria; hence, the methods will yield differing results with an increase in false positive results (Fiksdal and Tryland 2008, Davies et al. 1994, Davies et al. 1995, Fiksdal and Tryland 1999, Pommepuy et al. 1996, Tryland et al. 1998).

Today's waterworks must comply with their countries' current regulations based on growth methods. Lower levels of coliforms can easily be detected by simply extending the incubation time, such as the growth methods in Colifast CALM at 12 hours.

3.2.2 BACTcontrol

In February 2014, microLAN acquired mbOnline and added the Coliguard system to their product range. Coliguard was renamed and is now known as BACTcontrol. Possible applications are untreated water, drinking water and waste water. The BACTcontrol is a system for measuring the number of fecal coliforms in the field and is based on fluorescent measurement of the end-product of specific enzymatic activity with no growth phase in the measuring cycle. The system consists of pumps and valves for liquid handling, a filtration unit connected to a heater and temperature controller and a detector unit. The analysis, initiated by adding buffers and substrate solutions to the reaction chamber, is controlled by a data and control unit. Depending on the sample volume, ranging from 100 to 5000 ml, the instrument is claimed to detect down to 1–5 cfu/100 ml (Bjergaarde and Hansson 2012). For clean water applications, with low levels of bacteria, the detection time is 3–5 hours with 3000–5000 ml sample volume. The system concentrates bacteria from the sample using a ceramic filter prior to addition of substrate and analysis. The filter is reused after rinsing and can be used for 75 to 450 measuring cycles, depending on the turbidity. For waste water applications, with high levels of bacteria, the system can be used without filtration and the detection time is reduced to < one hour. The normal measuring cycle takes 2 to 3 hours, with an express cycle taking one hour. The duration of the filtration phase depends upon volume and control condition of the filter and the measurement cycle takes approximately 60 minutes and consists of heating the incubator to 36/44 degrees C, adding reagents, measurement of fluorescent light and calculation of the bacteria level (Fig. 9). The instrument cleans itself by heating and sterilization with rinse solutions between analyses. Results can be transferred to an industrial SCADA system. The advantages are—the automated rapid results, the differentiation between *E. coli*, Coliform and total bacteria activity and the

Figure 9. The microLAN BACTcontrol system including an integrated touchscreen PC for handling and transferring analysis data.

disadvantages are the possible non-growth unspecific results (compared to laboratory results using traditional cultivation methods) and some technical challenges related to filtration (microLAN, Netherlands).

3.2.3 ColiMinder

The ColiMinder was launched by Vienna Water Monitoring in 2013. The technology is based on detection of fluorescence and is a rapid enzymatic method without bacterial growth. The system comes in three versions—one bench top analyzer (ColiMinder), one portable battery powered unit (ColiMinder Mobile) and one outdoor measuring station (ColiMinder OMS). All the three versions contain pumps and valves, an incubation and detection chamber and a data-and-control unit. ColiMinder detects what they refer to as total *E. coli*, which includes both viable and non culturable bacteria (VBNC). The ColiMinder detection unit is the enzymatic activity of beta-D-glucuronidase expressed in Modified Fishman Units per 100 ml of water sample (MFU/100 ml). The MFU value is used to estimate the level of VBNC. Typical sample volume is 6–7 ml and the analysis is based on addition of a buffer solution and substrate solutions to the sample. The results are given in 10–30 minutes and the whole process, including cleaning with rinse solutions, is fulfilled in 40–60 minutes with the detection range being 0,001 to 1000 MFU/100 ml (Koschelnik 2015). The system can run multiple samples before restart and refill—from 10 samples with the portable version to >100 for the bench model. Results can be transferred by a network connection (Fig. 10). This system has the same advantages and disadvantages as the BACTcontrol system but it is not affected by challenges with an up-concentrating step prior to analysis (Vienna Water Monitoring, Austria).

Figure 10. The ColiMinder system integrated in cabinet as an outdoor measuring station.

3.2.4 Colifast CALM

The CALM system can run a rapid MUP (methylumbelliferone production) format. This system consists of two incubator blocks (80 positions/wells), a random-access robot, valves and pumps, a detector unit and an integrated computer with software (Fig. 11). The rapid 2-hour MUP test performed by the system includes automatic sampling of 10 ml samples, mixing with Colifast 6 growth medium in wells and incubation at 44°C. The medium includes substrates and the fluorescence (MU) produced by the bacterial enzyme activity is measured several times by sub-sampling from the test vials/wells. The measured increase in fluorescence is calculated by the system to ppbMU per hour and this number indicates the level of fecal coliforms in 100 ml. The system uses a rinse solution to clean out the components prior to the next analysis. The total sampling and measurement cycle takes 2.5 hours and the system can run 76 samples during one run. The same rapid-method advantages and disadvantages apply to this system. Due to greater variability and environmental interference at low bacterial levels, the Colifast rapid method is recommended on samples containing > 500 coliforms/100 ml (Colifast AS, Norway).

3.3 Molecular methods

In order to fulfil the constant need for more rapid methods for detection of indicators in the water industry and ideal real-time analysis, several

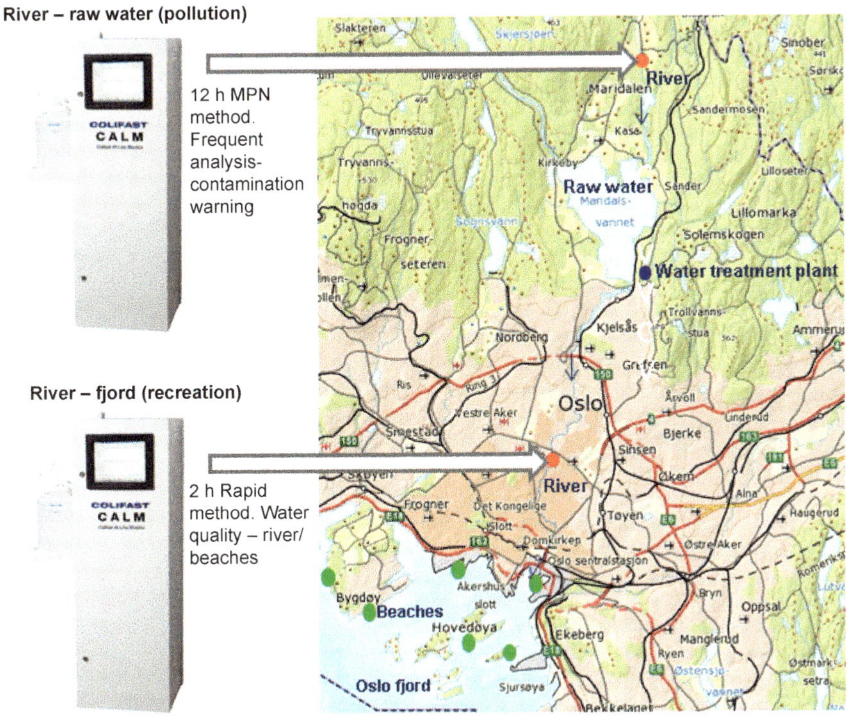

Figure 11. Rapid fecal contamination monitoring by CALM systems in Oslo, Norway. The large urban River Akerselva is occasionally contaminated by sour overflows during heavy rainfall, which affects the water quality of recreational areas downstream and at beaches in the fjord close to the river outlet. By use of the rapid methods the level of coliforms is analyzed with a high frequency and with results available 2.5 hours after sampling.

molecular methods have been proposed. These methods normally combine the concentration procedures, DNA enrichment steps and analysis with *E. coli* specific real-time PCR (Maheux et al. 2011). In recent years there is reported improved sensitivity for molecular methods and some of these can now be used for monitoring drinking water. However, the methods rely on complex laboratory procedures and are currently (2016) not integrated into any automated online systems.

3.3.1 Surrogate parameters

Surrogate parameters like adenosine triphosphate (ATP) and spectral absorption coefficients can give useful information about the water quality, though these are not specific for fecal contamination and will usually not have a good correlation with the number of fecal bacteria found by specific growth methods (Stadler et al. 2010, Hammnes et al. 2010). The previously described rapid enzyme activity methods proved to be somewhat comparable but less accurate compared to the growth methods. They can be classified as a more specific and suitable surrogate parameter for detecting fecal contamination.

4. Verification and Validation of New Technology

Verification and validation of automated rapid methods for detection of indicator bacteria in water must be a holistic process that includes instrumentation, software, consumables and reagents in addition to the specific detection. The process might be done by the customer herself, or by the producer or a qualified third party to assess emerging technologies with limited results, experience and references. The Environmental Technology Verification (ETV) of the EU Joint Research Center is a new tool to help innovative environmental technologies reach the market. Claims about the performance of innovative technologies can be verified and a similar program was formerly conducted by the US EPA. The Colifast ALARM received the first verification statement for an at-line monitor for fecal contamination from the US EPA ETV in 2011 (EPA 2011).

5. Future Technologies and Recommendations

To further reduce the time for action, future methods will avoid time-consuming growth of indicator bacteria and focus on other means of identification. Efforts will be made to find a target molecule that does not remain stable after the cell death and methods that are able to distinguish between viable and non-viable cells. New methods should be sensitive and able to detect one target cell in a specified volume of water and be specific and able to detect the target molecule from a complex water sample with a background of non-target microorganisms. Furthermore, the method must be taken out of the laboratory into the automated sampling and testing devices. It is possible that a higher recovery of target organisms may occur. Then the current legislation will have to be altered in order to incorporate the new technology.

6. Conclusion

The emergence of automated fecal monitoring systems for water started in the mid 90s and during the past 20 years, several systems have been introduced in the market. Some of these disappeared after a short time while others were more successful and are still available. The technical designs are different but some of the methodological principles are quite similar. All the known fully-automated fecal bacteria monitors today utilize specific enzymatic activity of the target bacteria with corresponding substrates. The different detection formats are based on both bacterial growth, in combination with bacterial enzyme activity, and rapid non-growth detection using only sensitive measurements of the initial activity. The growth methods are known to generate accurate results and usually show a good correlation with the standard reference methods while the more rapid enzymatic methods have shown to be less accurate. While the rapid methods generate a result in 0.5 to 2 hours, the automated growth methods have a typical analysis time of 12–17 hours. The higher quality of the growth methods is due to 3 selective steps for detection:

- Target bacteria have the ability to multiply/grow
- The increasing number of bacteria correspondingly increases the enzyme concentration
- The sample will be classified positive only when it shows a high level of end-product from enzymatic reaction.

These steps ensure accurate detection of viable bacteria, and the high level detection limit discriminates interference from other factors in the sample. These methods are also more similar in design to the traditional growth method used in laboratories. In contrast, the rapid methods have only one step to ensure accuracy; the specific enzyme activity. And this will usually only be at very low concentrations and without access to include an elevated limit to discriminate environmental interference. The contents in different water samples can be complex and many samples have a variety of unspecific enzyme activity from dead target bacteria, free enzymes, algae and some non-target bacterial strains with similar activity (Baudart et al. 2009). External factors, like sunlight, pre-treated pollution and temperature can also affect the rapid results. The limitations and advantages must be considered when selecting a format and an appropriate system for a specific application. Based on the use of the available automated systems today, the automated growth method formats are best suited when comparability with results from laboratories is important. These systems are used in analyzing drinking water, recreational waters, rivers and lakes and waste water. The rapid enzymatic methods can be used when the accuracy is less important and these systems are best suited when the target bacterial levels are medium to high due to less background interference at these levels. A typical application would be environmental monitoring. The technical quality and stability of the different systems may vary. Documentation, including verification statements and test reports, are available for some of the systems. Technical stability and durance information may be available through references and system operators, e.g., one of these systems, the

Colifast CALM, is reported to operate for more than 14 years (2017) without technical problems. Since the year 2000 till today, the number of analyzers in use has slowly been growing, which indicates an increasing interest in the market for automated bacterial monitors. As these systems successfully monitor relevant fecal indicators, decrease the time for the results, increase the test frequency and replace cost and time-consuming manual analysis, the demand for this automated technology is likely to increase in the future.

Keywords: *E. coli*, thermotolerant coliforms, fecal coliforms (FC), MPN (most probable number), TTD (time to detect), enzyme activity, PA (presence/absence) format

References

American Public Health Association (APHA). 2005. American Water Works Association. Water Environment federation. Standard Methods for the Examination of Water and Wastewater. 21st ed.; Washington, DC, USA.

Apte, S.C., Davies, C.M. and Peterson, S.M. 1995. Rapid detection offaecal coliforms in sewage using a colorimetric assay of β-Dgalactosidase. Australia Water Research 07/1995; 29(7): 1803–1806.

Baudart, J., Servais, P., De Paoli, H., Henry, A. and Lebaron, P. 2009. Rapid enumeration of *Escherichia coli* in marine bathing waters: Potential interference of nontarget bacteria. J. Appl. Microbiol. 107: 2054–2062.

Berg, J.D. and Fiksdal, L. 1988. Rapid detection of total and fecal coliforms in water by enzymatic-hydrolysis of 4-methylumbelliferone-beta-D-galactoside. Appl. Environ. Microb. 54: 2118–2122.

Berg, J.D. 1991, 1993. Patent number: 5292644, Rapid process for detection coliform bacteria, 1991. Patent number: 5518894, Rapid coliform detection system, 1993.

Bjergaarde, N.E. and Hansson, E. 2012. New methods for on-line detection of coliforme bacteria in drinking water. Åttonde Nordiska Dricksvattenkonferensen 105–114.

Braathen, H., Ranneklev, S., Sagstad, H. and Rydberg, H. 2005. Fully automated monitor reduces intake of *E. coli*-contaminated raw water. Water and Wastewater International. April 2005, 47.

Cabral, João, P.S. 2010. Water microbiology. Bacterial pathogens and water. Int. J. Environ. Res. Public Health 7: 3657–3703.

Centers for Disease Control and Prevention (CDC). 2015. Morbidity and Mortality Weekly Report (MMWR). Surveillance for Waterborne Disease Outbreaks Associated with Drinking Water. United States, 2011–2012.

Davies, C.M., Apte, S.C. and Peterson, S.M. 1995. Beta-D-galactosidase activity of viable, non-culturable coliform bacteria in marine waters. Letters in Applied Microbiology 21: 99–102.

Davies, C.M., Apte, S.C., Peterson, S.M. and Stauber, J.L. 1994. Plant and algal interference in bacterial beta-D-galactosidase and beta-D-glucuronidase assays. Appl. Environ. Microb. 60: 3959–3964.

Demowatercoli Final Report. EC Framework 5. 2003. [(accessed on 22. december 2015)]. Available online: http://www.colifast.no/wp-content/uploads/pdf/EU%20Demoatercoli.pdf.

Eckner, K.F., Jullien, S., Samset, I.D. and Berg, J.D. 1999. Rapid, enzyme-based, fluorometric detection of total and thermotolerant coliform bacteria in water samples. Water Supply-Oxford. Rapid microbiological monitoring methods von Blackwell Science 17(2): 73–80.

Edberg, S.C., Allen, M.J. and Smith, D.B. 1988. National field-evaluation of a defined substrate method for the simultaneous enumeration of total coliforms and *Escherichia coli* from drinking-water—comparison with the standard multiple tube fermentation method. Appl. Environ. Microb. 54: 1595–1601.

Edberg, S.C., Allen M.J., Smith, D.B. and Kriz, N.J. 1990. Enumeration of total coliforms and *Escherichia coli* from source water by the defined substrate technology. Appl. Environ. Microbiol. 1990 Feb; 56(2): 366–369.

Environmental Protection Agency (EPA). 2011. Environmental Technology Verification Report, Colifast Alarm at-line automated remote monitor. Available online: http://nepis.epa.gov/Adobe/PDF/P100BSVR.pdf.

European Union (EU). 1998. Council Directive 98/83/EC of 3 November 1998 on the quality of water intended for human consumption.

European Union (EU). 2006. Directive 2006/7/EC of the European Parliament and of the council of 15 February 2006 concerning the management of bathing water quality and repealing directive 76/160/eec.

Fiksdal, L., Pommepuy, M., Caprais, M.P. and Midttun, I. 1994. Monitoring of faecal pollution in coastal waters by use of rapid enzymatic techniques. Appl. Environ. Microbiol. 60: 1581–1584.

Fiksdal, L. and Tryland, I. 1999. Effect of U.V. light irradiation, starvation and heat on *Escherichia coli* beta-D-galactosidase activity and other potential viability parameters. J. Appl. Microbiol. 87: 62–71.

Fiksdal, L. and Tryland, I. 2008. Application of rapid enzyme assay techniques for monitoring of microbial water quality. Curr. Opin. Biotech. 19: 289–294.

George, I., Petit, M. and Servais, P. 2000. Use of enzymatic methods for rapid enumeration of coliforms in freshwater. J. Appl. Microbiol. 88: 404–13.

Hammnes, F., Goldschmidt, F., Vital, M., Wang, Y.Y. and Egli, T. 2010. Measurement and interpretation of microbial adenosine tri-phosphate (ATP) in aquatic environments. Water Res. 44: 3915–3923.

Koschelnik, J., Vogl, W., Epp, M. and Lackner, M. 2015. Rapid analysis of β-D-glucuronidase activity in water using fully Automated Technology. Water Resources Management, 471–482.

Maheux, A.F., Bissonnette, L., Boissinot, M., Bernier, J.L.T., Huppe, V., Picard, F.J., Berube, E. and Bergeron, M.G. 2011. Rapid concentration and molecular enrichment approach for sensitive detection of *Escherichia coli* and Shigella species in potable water samples. Appl. Environ. Microb. 77: 6199–6207.

Pommepuy, M., Fiksdal, L., Gourmelon, M., Melikechi, H., Caprais, M.P., Cormier, M. and Colwell, R.R. 1996. Effect of seawater on *Escherichia coli* beta-galactosidase activity. J. Appl. Bacteriol. 81: 174–180.

Ryzinska-Paier, G., Lendenfeld, T., Correa, K., Stadler, P., Blaschke, A.P., Mach, R.L., Stadler, H., Kirschner, A.K.T. and Farnleitner, A.H. 2014. A sensitive and robust method for automated on-line monitoring of enzymatic activities in water and water resources. Water Sci. Technol. 69: 1349–1358.

Stadler, H., Klock, E., Skritek, P., Mach, R.L., Zerobin, W. and Farnleitner, A.H. 2010. The spectral absorption coefficient at 254 nm as a real-time early-warning proxy for detecting faecal pollution events at alphine karst water resources. Water Sci. Technol. 62: 1898–1906.

Tryland, I., Pommepuy, M. and Fiksdal, L. 1998. Effect of chlorination on beta-D-galactosidase activity of sewage bacteria and *Escherichia coli*. J. Appl. Microbiol. 85: 51–60.

Tryland, I., Samset, I.D., Hermansen, L., Berg, J.D. and Rydberg, H. 2001. Early warning of faecal contamination of water: A dual mode, automated system for high-(< 1 hour) and low-levels (6–11 hours). Water Sci. Technol. 43: 217–22.

Tryland, I., Surman, S. and Berg, J.D. 2002. Monitoring faecal contamination of the Thames estuary using a semiautomated early warning system. Water Sci. Technol. 46(3): 25–31.

Tryland, I., Eregno, F.E., Braathen, H., Khalaf, G., Sjolander, I. and Fossum, M. 2015. On-line monitoring of *Escherichia coli* in raw water at oset drinking water treatment plant, Oslo (Norway). Int. J. Env. Res. Pub. He. 12: 1788–1802.

United Nations (UN). 2010. The human right to water and sanitation, Resolution 64/292.

Van Poucke, S.O. and Nelis, H.J. 1995. Development of a sensitive chemiluminometric assay for the detection of beta-galactosidase in permeabilized coliform bacteria and comparison with fluorometry and colorimetry. Appl. Environ. Microbiol. 1995 Dec; 61(12): 4505–4509.

World Health Organization (WHO). 2001.Water quality: Guidelines, standards and health. Chapter 13. Indicators of microbial water quality.

World Health Organisation (WHO). 2011. Guidelines for drinking water treatment. 4th ed. Microbial aspects. World Health Organisation; Geneva, Switzerland.

Zuckerman, U., Hart, I. and Armon, R. 2008. Field evaluation of colilert 3000 for ground, raw and treated surface water and comparison with standard membrane filtration method. Water Air and Soil Pollution 188: 3–8.

2

Automated Near-real-time Monitoring of Enzymatic Activities in Water Resources

P. Stadler,[1,2,*] *G. Ryzinska-Paier,*[3] *T. Lendenfeld,*[4] *W. Vogl,*[5] *A.P. Blaschke,*[6,7] *P. Strauss,*[8] *H. Stadler,*[9] *M. Lackner,*[10] *M. Zessner*[1,2] and *A.H. Farnleitner*[3,7]

1. Introduction

Sensitive and rapid detection of microbiological contaminants is essential for sustainable and proactive management of water resources. Because cultivation-based standard analyses of fecal pollution typically require more

[1] TU Wien, Centre for Water Resource Systems, A-1040, Vienna, Austria.
[2] TU Wien, Institute for Water Quality, Resources and Waste Management, Karlsplatz 13, A-1040 Vienna, Austria.
[3] TU Wien, Institute of Chemical Engineering, Research Group Environmental Microbiology and Molecular Ecology, Gumpendorferstraße 1a, 1060 Vienna, Austria.
[4] WSB Labor-GmbH, Steiner Landstraße 27a, 3500 Krems, Austria.
[5] Vienna Water Monitoring solutions, Dorfstrasse 17, A-2295 Zwerndorf, Austria.
[6] TU Wien, Institute of Hydraulic Engineering and Water Resources Management, E222/2, Karlsplatz 13 A-1040 Vienna, Austria.
[7] Interuniversity Cooperation Centre for Water and Health, TU Wien & Medical University of Vienna, Gumpendorferstraße 1a, 1060 Vienna, Austria; www.waterandhealth.at.
[8] Federal Agency for Water Management, Institute for Land & Water Management Research, 3252 Petzenkirchen, Austria.
[9] Joanneum Research, Institute for Water, Energy and Sustainability, Department of Water Resources Management, Elisabethstrasse 16/II, 8010 Graz, Austria.
[10] Institute of Advanced Engineering Technologies, University of Applied Sciences FH Technikum Wien, Höchstädtplatz 6, A-1200 Vienna, Austria.
* Corresponding author: philipp.stadler@gmx.at

than one working day, these methods are not suitable for rapid water quality assessment (Cabral 2010). Alternative methods involving enzymatic activity measurements have been tested in various aquatic habitats (Farnleitner et al. 2002, 2001, Fiksdal and Tryland 2008, Garcia-Armisen et al. 2005). Over the last two decades, several studies have suggested the use of direct enzymatic activity determination to monitor microbiological contamination in various water sources (Farnleitner et al. 2002, Fiksdal et al. 1994a, Fiksdal and Tryland 2008, George et al. 2001). These common enzymatic activity measurements for fecal indicators require laboratory facilities and elaborate sampling methods (Lebaron et al. 2005, Rompré et al. 2002) and studies have reported a tight association between measurements of enzymatic activity and culture-based analyses (Farnleitner et al. 2002, Garcia-Armisen et al. 2005, George et al. 2000, Ouattara et al. 2011), but these measurements remain too time-consuming for rapid application.

Up-to-date technological developments have enabled fully automated measurements of enzymatic activity (e.g., using glucuronidases or galactosidases) that are suitable for on-site and real-time monitoring (Koschelnik et al. 2015, Ryzinska-Paier et al. 2014, Zibuschka et al. 2010). Measurements were reported to be possible in less-than-hourly intervals (Koschelnik et al. 2015, Stadler et al. 2016). The potential for real-time monitoring of enzymatic activity seems to be high, specifically for implementation into early warning systems and process control. Moreover, this method will help to improve the understanding of the catchment behavior as well as contaminant transport processes in different habitats. In this respect, there is a need for studies to test prototypes for automated and on-site enzymatic activity determination to evaluate the consistency of measurement results, technical robustness during long-term on-site operation and proxy capability for culture-based microbiological analyses in the observed habitat.

These specific questions were already addressed in recent research by Ryzinska-Paier et al. 2014 and Stadler et al. 2016, focusing on prototype operations either in groundwater (Ryzinska-Paier et al. 2014) or in surface water (Stadler et al. 2016). The test sites of these studies differ in hydrogeology, microbiological impact and hydrological catchment dynamics. However, till now, no scientific overview exists that compares the application of automated measurements in these contrasting environments. The aim of this chapter is: (i) to evaluate whether technical devices for automated measurements of enzymatic activity are technically robust for long-term on-site operation at groundwater and surface water monitoring-locations, (ii) whether automated GLUC determination can be used as a proxy for culture-based *E. coli* analyses in various habitats and (iii) whether enzymatic GLUC activity measurements can be used as a general indicator for the fecal contamination of water resources. In addition, further research needs regarding this innovative technology are discussed.

The experimental results from the aforementioned studies were used in this chapter to describe the application of on-site enzymatic methods. This chapter's objective is to assess from an engineering point of view, the question of whether automated measurements of enzymatic activity are possible in differing and technically challenging aquatic habitats. Furthermore, aspects of the proxy capability of on-site measured GLUC activity for culture-based analyses and the indicator potential of GLUC activity for fecal contamination of water are addressed.

2. Materials and Methods

2.1 Data

The basis for this work emerges from research published by Ryzinska-Paier et al. (2014) and Stadler et al. (2016) describing the realization of automated on-site measurements of beta-D-glucuronidase in groundwater (Ryzinska-Paier et al. 2014) and surface water (Stadler et al. 2016). The authors chose these studies as excellent sources of data to comparatively review the application of these measurements in contrasting water resources, including surface and groundwater. The following key characteristics of the studies support their comparison: (i) both studies were conducted independently on test sites but provide a detailed background description of the aquifer characteristics, microbiological impact range and dynamics; and (ii) the tested prototypes used the same substrate for the specific determination of beta-D-glucuronidase (GLUC) activity. Furthermore, data from laboratory analyses of GLUC activity are available in the literature (Farnleitner et al. 2002, Garcia-Armisen et al. 2005, George et al. 2000, Ouattara et al. 2011), and therefore, a comparison between on-site and laboratory assays is appropriate.

2.2 Test sites

2.2.1 Karstic limestone aquifer

The karst spring LKAS2 is located in the Northern Calcareous Alps in Austria (Fig. 1). The hydrogeological catchment reaches altitudes of approx.

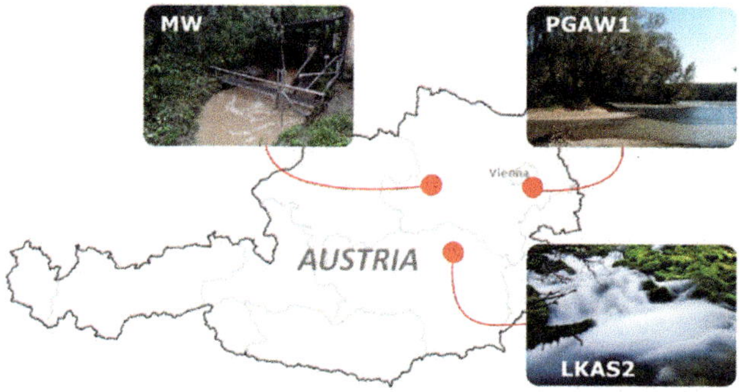

Figure 1. Map of Austria. Test sites LKAS2, PGAW1 and MW are marked with red dots.

2200 m, with a total area of approx. 60 km². The hydrological regime is highly dynamic (Table 1) and is characterized by a prompt hydraulic response to intense precipitation events (Stadler et al. 2008, 2010). During the test period, a discharge of almost 30 $m^3 s^{-1}$ and a minimum of 1.1 $m^3 s^{-1}$ was recorded. The water temperature is fairly constant over the year and the median of 5.4°C represents an alpine catchment in this geographic region. Turbidity values reached a maximum of 4.5 FNU (Formazine Nephelometric Units) during intense runoff events. LKAS2 is highly vulnerable and the dominant source of fecal pollution (ruminant) is surface-associated input during the grazing period in summer due to hydrologic events (Reischer et al. 2008).

2.2.2 Porous groundwater aquifer

The well PGAW1 (Fig. 1) is located in an alluvial backwater area downstream of Vienna (Vierheilig et al. 2013). The water quality is stable throughout the year and the impact of fecal contamination is not present widely. Contamination only tends to occur during flood events of the Danube river (Kirschner et al. 2014). The physico-chemical parameters (Table 1) coincide with a well assigned to a porous groundwater aquifer.

2.2.3 Surface water

MW is a monitoring location at a stream discharging a 0.66 km² catchment with 87 per cent agricultural land use. The HOAL (Hydrological Open Air Laboratory, Blöschl et al. 2015) catchment is an experimental catchment in western Lower Austria (Fig. 1). It is characterized by elevated discharge dynamics (Table 1) with a rapid response to precipitation events (Exner-

Table 1. Comparative overview of the ranges of the measured basic hydrological parameters and fecal pollution by the cultivation-based standard fecal indicator *E. coli* at the three test sites (n = number of measurements, FNU = Formazine Nephelometric Units, MPN = Most Probable Number).

	LKAS2 (2010–2011)				**PGWA1 (2010–2011)**				**MW (2014–2015)**			
	n	median	min	max	*n*	median	min	max	*n*	median	min	max
Discharge (LKAS2, MW) Abstraction rate (PGWA1) [l/s]	*99423*	4107	1112	29795	*7589*	65	0	262	*8760*	2.3	0.5	73.4
Turbidity [FNU]	*110368*	0.3	0.0	4.5	*101370*	0.2	0.1	0.9	*8760*	8	0	3210
Conductivity [µS/cm]	*105013*	195	159	222	*11904*	563	493	650	*8760*	769	260	856
Temperature [°C]	*104933*	5.4	4.9	5.9	*13449*	11.4	9.8	13.1	*8760*	10.7	0.2	20
E. coli [MPN/100 ml]	*206*	0	0	435	*214*	0	0	0	*54*	172	0	3450

Kittridge et al. 2013). During the test phase, a discharge maximum of 73 l/s and a minimum of 0.5 l/s was recorded. During event runoff conditions, the stream-water is sediment-laden, as indicated by turbidity values of up to 3210 FNU. The stream-water temperature ranged between 0.2°C and 20°C following the trend of the yearly air temperature. The main source of fecal contamination is manure applied periodically to crop fields (Stadler et al. 2016).

2.3 On-site GLUC measurements

For automated enzymatic activity determination, prototypes of two different devices were tested—BACTcontrol, formerly Coliguard: MicroLan, Netherlands; ColiMinder: VWM, Austria. Both the designs detect beta-D-glucuronidase enzymatic activity and record and transmit the data in near-real-time. The measurement principle is based on a photometric measurement chamber that enables high-resolution fluorescence analysis. During the measurement process, the sample mixed with specific assay reagents (proprietary information) generates an increasing fluorescence signal, reflecting the level of enzymatic activity, which is monitored over time. Internal control parameters, such as the fluorescence signal, linearity of the fluorescence slope, temperature of the measurement chamber, the device's environmental temperature, the measurement duration and blank value measurements are available for each data point and were used initially to quality check the measurement results. All devices were connected to a GPRS modem for data transfer and on-line access to the measurement device. GLUC activity measurements from both devices were performed in batches. ColiMinder devices used 6.5 ml of sample per measurement. The full ColiMinder measurement cycle, including cleaning and sample conditioning, lasts 30 to 40 minutes. BACTcontrol devices (Ryzinska-Paier et al. 2014, Zibuschka et al. 2010) enable adjustable sampling volumes (100 ml up to 5000 ml). For ground and spring water monitoring, 1000 ml was used; for surface water, 100 ml sample volume was used. The full BACTcontrol measurement cycle, including cleaning and sample conditioning, lasts 180 minutes. ColiMinder is calibrated to Modified Fishman Units (MFU/100 ml), based on the enzyme unit definition for beta-glucuronidase activity (Fishman and Bergmeyer 1974, Bergmeyer 2012). BACTcontrol provides units of pmol/min/100 ml. Stadler et al. (2016) suggest an average one-to-one ratio between mMFU/100 ml and pmol/min/100 ml. Further technical details and information about the comparison of BACTcontrol and ColiMinder can be found in Stadler et al. 2016.

At test sites PGAW1 and LKAS2, BACTcontrol devices (MicroLan, Netherlands) for automated beta-D-glucuronidase (GLUC) activity were operated over a two-year period (2010 to 2011). At location MW, both BACTcontrol and ColiMinder were operated in parallel for over a year (2014 to 2015).

On-site-measured GLUC data from all the three test sites were compared to physico-chemical parameters monitored in parallel. The GLUC signals gathered at location MW with devices having two different constructions were compared with each other by performing linear regression analysis.

2.4 Hydrological and microbiological parameters

E. coli was used as a microbiological standard parameter of fecal pollution (Farnleitner et al. 2010). Water samples were analyzed for the cultivation-based bacterial standard fecal indicator *E. coli* using Colilert-18 (ISO 9308-2:2012, MPN/100 ml). The sampling intervals ranged from biweekly (PGAW1, LKAS2) to monthly (MW).

At all test sites, on-line sensors were used to gather hydrological as well as physico-chemical parameters. On-site measurements of discharge (Q), electrical conductivity (EC), water temperature and turbidity in high temporal resolution enabled ascertainment of the hydrological conditions in the observed catchment (Ryzinska-Paier et al. 2014, Stadler et al. 2016). The captured hydrologic dynamics, characteristic for each test-site, were used to assess the plausibility of automated GLUC measurements (e.g., event runoff conditions determined by decreasing the electrical conductivity and increasing the turbidity as an indicator of the potential influence of contaminated event water). The data from environmental background parameters were compared to GLUC measurements using Spearman's rank correlation.

2.5 Comparison with standard laboratory enzymatic assays

Ryzinska et al. (2014) performed laboratory experiments to assess the comparability of automated enzymatic activity measurements (BACTcontrol) and standard enzymatic assays. A Sigma Aldrich assay was applied and methylumbelliferyl-β-D-glucuronide (MUG) was used as a substrate. Two comparative analyses were conducted, ranging from a short incubation time up to 75 minutes of incubation. The release of methylumbelliferyl (MUF) was evaluated using high performance liquid chromatography (HPLC).

3. Results

3.1 Technical realization

3.1.1 Technical application and long-term on-site operation

The tested prototypes proved to be reliable for continuous long-term operation of enzymatic measurements at the various test sites (Table 2). Measuring and cleaning procedures were conducted automatically and the

Table 2. Comparative data for the tested devices at the contrasting surface and groundwater locations (modified from Stadler et al. 2016 and Ryzinska-Paier et al. 2014).

	ColiMinder	**BACTcontrol**
Test-site	*MW*	*PGAW1, LKAS2, MW*
Company	Vienna Water Monitoring (Austria)	MicroLan (Netherlands)
Tested substrate	beta-d-glucuronidase (GLUC)	beta-d-glucuronidase (GLUC)
Parameter	mMFU/100 ml	pmol/min/100 ml
Limit of quantification	0.8 mMFU/100 ml	1.5 pmol/min/100 ml
Time resolution (measurement incl. cleaning cycle)	60 min	180 min
Data transfer	GPRS modem	GPRS modem
Internal control parameters (metadata)	Fluorescence signal, slope of signal, temperature (measuring chamber, device), measurement duration, blank value measurement	Fluorescence signal, slope of signal, temperature (measuring chamber, device, LED), measurement duration, pump rating, blank value measurement
Blank value measurement (programmable)	every 12 hours	every 24 hours
Total test time	*MW*: 12 months	*MW*: 6 months* *PGAW1, LKAS2*: 12 months
2 devices operated in parallel	*MW*: ColiMinder-01, ColiMinder-02	*MW*: BACTcontrol-01, BACT control-02
Regular service interval (e.g., reagent refill, cleaning of tubing and filter)	*MW*: biweekly	*MW*: biweekly *PGAW1, LKAS2*: monthly
Technical service (e.g., re-calibration)	3–6 months	6–12 months

* *BACTcontrol devices were operated at MW since 2012, but in this study, only measurements after the installation of an improved sampling setup in July 2014 are used.*

measurement results were transmitted using the GPRS network. However, considerably shorter intervals for manual maintenance were needed for surface water operation as compared to those reported for groundwater (Ryzinska-Paier et al. 2014, Stadler et al. 2016). On-site measurement data could be gathered for up to 6 months without technical failure at all the locations, even in the presence of a total suspended solid concentration of up to 3 g/l (MW).

At pristine groundwater resources (Table 1, Table 2), such as site PGWA1, the level of GLUC was below the detection threshold of the assay. At sediment-laden stream water (MW), the main technical challenge was the high suspended solid (TSS) load in the monitored water during event runoff conditions. In this case, on-site sample pre-filtration was necessary to prevent the tubing and valves from clogging (Stadler et al. 2016). Reference analytics of unfiltered and filtered water samples showed no significant effect of filtration through 100 μm pore size on *E. coli* concentrations and enzymatic activity (Stadler et al. 2016).

3.1.2 Comparability of automated GLUC measurements between devices

At location MW, four prototypes having two different constructions (2x BACTcontrol, 2x ColiMinder) were operated in parallel (Stadler et al. 2016). Linear regression analyses (all p-values <0.001) showed highly consistent results for devices with the same construction. A linear correlation coefficient, R^2, of 0.94 was found between the two ColiMinder apparatuses and an R^2 value of 0.96 was obtained for the two BACTcontrol devices (Stadler et al. 2016). The correlations between devices with different designs showed reasonable consistency, with an R^2 value of 0.71 (Fig. 2).

BACTcontrol and ColiMinder drew samples from opposite sides of the stream and sampling times may have differed by up to one hour, which very likely contributed to the lower correlation. Overall, measurements from the two designs had a highly symmetrical range and an average one-to-one ratio of signals (Stadler et al. 2016).

3.1.3 Comparability to laboratory standard enzymatic assays

For both the tested incubation times, the automated enzymatic measurements yielded results that were highly comparable to a standard Sigma assay (Ryzinska-Paier et al. 2014).

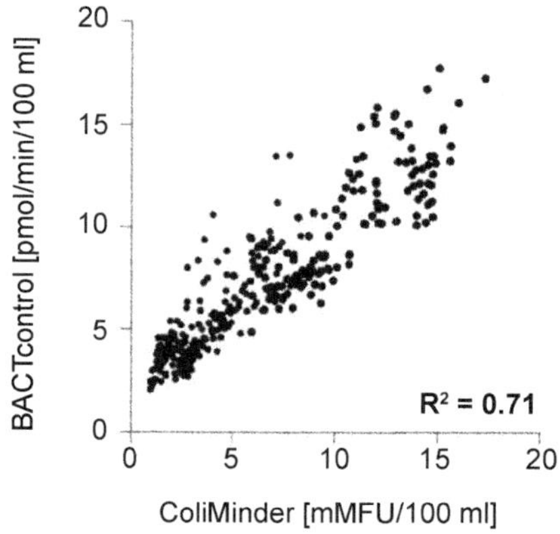

Figure 2. Scatter plot showing the correlation of GLUC measurements conducted with devices having different construction (modified from Stadler et al. 2016).

3.2 Range and dynamics of GLUC signals

GLUC signals monitored on-site in high temporal resolution had fundamentally different ranges and dynamics of enzymatic activity at the various test sites. GLUC data ranged from the limit of detection (PGAW1) to considerable levels of enzymatic activity (MW). The dynamics of GLUC activity were in general related to the changing hydrological conditions in the respective catchment, as indicated through the physico-chemical and hydrological parameters measured in parallel.

At the porous alluvial ground-water resource (PGAW1), GLUC activity showed almost no variation (Fig. 3), ranging throughout the year within the limit of detection (Ryzinska-Paier et al. 2014).

At LKAS2, the GLUC signal reflected the very dynamic nature of a limestone-karstic aquifer (Fig. 3). In particular, events of intense summer precipitation caused peaks of enzymatic activity, reaching up to 6.0 pmol/min/100 ml. In winter, one flooding event caused an increase of GLUC values up to 2.5 pmol/min/100 ml (Ryzinska-Paier et al. 2014).

The GLUC activity at the stream monitoring location MW was highly dynamic (Fig. 3). A characteristic of this test site was the increased background of GLUC activity during the summer months. The base signal of GLUC activity reached its maximum in August/September, decreased

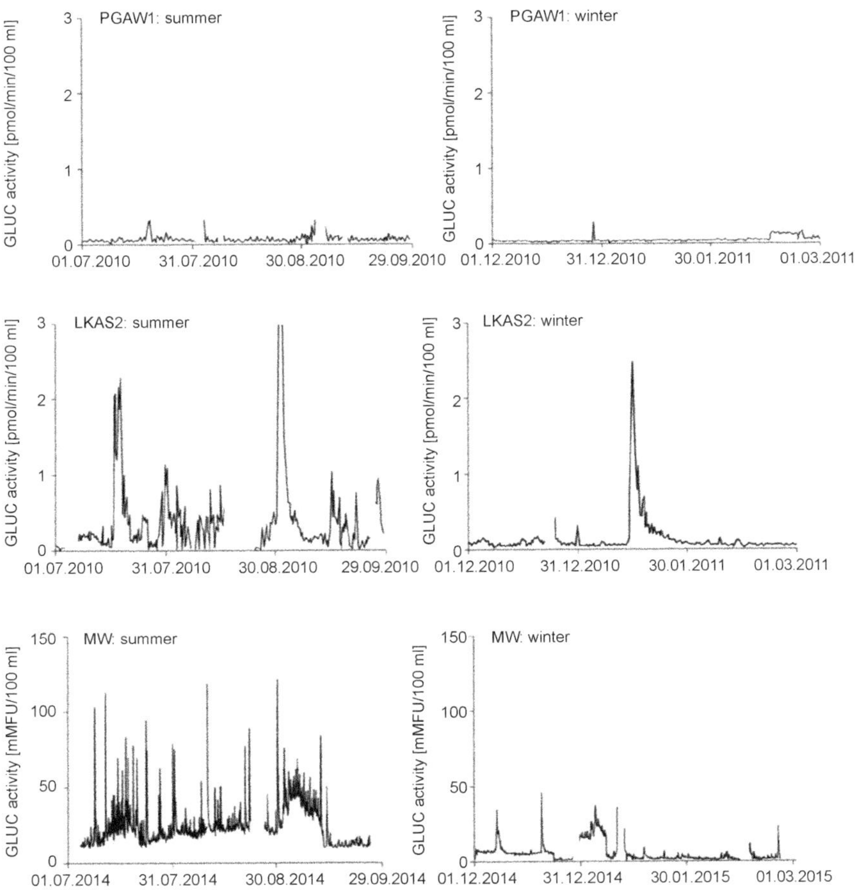

Figure 3. Graphic showing on-site measured GLUC activity at the test sites. Left plot shows summer months; right plot shows measurements during winter (plot for PGAW1 and LKAS2 reproduced from Ryzinska-Paier et al. (2014) with permission from the copyright holders, IWA Publishing, plot for MW modified from Stadler et al. (2016)).

during autumn and had a minimum in late winter (Fig. 3). Measurements taken with BACTcontrol ranged from 1.1 pmol/min/100 ml up to 108 pmol/min/100 ml (not shown here). ColiMinder recorded a minimum of 0.8 mMFU/100 ml and a maximum of 120 mMFU/100 ml. Precipitation events caused significant peaks of enzymatic activity in the stream water (Fig. 3). While all of the recorded hydrological events caused a prompt increase in GLUC activity in stream-water, the amplitude of GLUC peaks was not exclusively determined by the intensity of the event, stream-water turbidity or by the discharge (Stadler et al. 2016).

3.3 Comparison of GLUC data with hydrological and microbiological parameters

Cultivation-based *E. coli* were not detected in 100 ml samples and the automated GLUC values did not exceed the limit of quantification at site PGAW1 throughout the test period.

At location LKAS2, the cultivation-based *E. coli* concentrations ranged from undetectable to 435 MPN/100 ml. Analyses of cultivation-based *E. coli* from the stream-water samples collected at location MW revealed a (Table 3) of cultivation-based *E. coli* with isochronal on-site GLUC measurements showed a ρ of 0.53 at site LKAS2 (Ryzinska-Paier et al. 2014). At location MW, the correlation between *E. coli* and GLUC values was higher. Here, ρ of 0.71 for BACTcontrol measurements and ρ of 0.83 for ColiMinder measurements were found (Table 3).

At LKAS1, the GLUC measurements were more strongly correlated with the physico-chemical parameters, such as discharge (ρ=0.77) than with cultivation-based *E. coli*. At location MW, the GLUC signals gathered with both constructions were more tightly correlated with cultivation-based *E. coli*. Still, the GLUC measurements conducted with ColiMinder were more tightly correlated with the physico-chemical parameters than those conducted with BACTcontrol. This may be interpreted as an effect of the higher temporal resolution of ColiMinder measurements, allowing a clearer

Table 3. Comparison of GLUC measurements with environmental hydrological parameters and cultivation-based *E. coli* enumeration (ρ=Spearman's rank correlation coefficient, n=number of samples, P=p-value).

	LKAS2 BACTcontrol			**MW** BACTcontrol			**MW** ColiMinder		
	n	ρ	*P*	*n*	ρ	*P*	*n*	ρ	*P*
GLUC vs. discharge	*1804*	**0.77**	*<0.001*	*846*	**0.08**	*<0.05*	*1564*	**0.17**	*<0.001*
GLUC vs. turbidity	*1804*	**0.69**	*<0.001*	*846*	**0.35**	*<0.001*	*1564*	**0.72**	*<0.001*
GLUC vs. water temperature				*846*	**0.43**	*<0.001*	*1564*	**0.35**	*<0.001*
GLUC vs. *E. coli*	*113*	**0.53**	*<0.001*	*52*	**0.71**	*<0.001*	*52*	**0.83**	*<0.001*

tracking of promptly changing hydrological conditions (e.g., precipitation events). The correlations of GLUC values with the water temperature at MW revealed comparable levels for both BACTcontrol ($\rho = 0.43$) and ColiMinder ($\rho = 0.35$).

4. Discussion

4.1 Technical application

The tested devices for automated enzymatic activity determination proved to be reliable and robust under the diverse sets of field conditions they were subjected to. The time and staff required for the regular maintenance of the apparatuses was manageable and reasonable for on-site devices that are continuously operated, even for the monitoring of sediment-laden stream-water. These biweekly to monthly service intervals appear realistic for potential monitoring applications, such as the surveillance of process systems, wastewater treatment plants or bathing water.

Reference analyses with a standard enzymatic assay showed a high level of agreement between the results from automated devices and laboratory analyses. The comparison of different constructions of devices for the automated measurement of enzymatic activity showed comparable results.

The GLUC signals gathered at the various test sites appeared to be plausible with respect to the fact that they generally reflected the hydrological dynamics and catchment characteristics (Fig. 3). Furthermore, the ranges of GLUC activity determined for the examined test sites appeared reasonable when compared to each other and to previously published results (for further details, see 4.2).

4.2 Range and dynamics of GLUC signals

Previous studies reported GLUC values of up to 10^6 pmol/min/100 ml in water heavily influenced by municipal sewage (Farnleitner et al. 2002, Garcia-Armisen et al. 2005, George et al. 2000, Ouattara et al. 2011). The GLUC signals at station MW were consistent with an agricultural catchment subject to periodic manure application on the crop fields, ranging from the lower limit of detection to a value of signifying fecal contamination with maximum values of 108 pmol/min/100 ml and 120 mMFU/100 ml. The values for GLUC activity in fairly unpolluted karst water at site LKAS2 reached maximum values of 6 pmol/min/100 ml. At a protected groundwater resource, such as PGWA1, the values were below the limit of detection. On-site GLUC measurements at MW reached values almost twenty times higher than those at LKAS2. While the GLUC dynamics at the karst water resource LKAS2 were event-driven, the GLUC signal captured at MW was not exclusively determined by the hydrologic conditions. An increase in the baseline enzymatic activity signal during the summer was interpreted as the influence of ambient and water temperatures, as well as land management practices, on enzymatic activity.

4.3 GLUC as a proxy indicator for cultivation-based *E. coli* enumeration?

The evaluation of data gathered at LKAS2 and MW showed significant differences in the association between GLUC data and *E. coli* analyses by means of culture-based enumeration. Therefore, the eligibility of automated-measured GLUC activity as a proxy for standard microbiological methods was not verified. These results are in contrast to studies testing the comparability of laboratory enzymatic assays and culture-based analyses, where high associations between fecal indicator bacteria and GLUC were reported (Farnleitner et al. 2002, 2001, Fiksdal et al. 1994b, George et al. 2001). The discrepancy between results from GLUC enzymatic measurements and cultivation-based *E. coli* analyses, and the dependence of this association on the observed habitat, was demonstrated for the first time by the automated and on-site gathered GLUC measurements in different investigated catchments. Studies reporting tight correlations between GLUC and *E. coli* focused on catchments with a recent point-source influence from municipal sewage effluents (Farnleitner et al. 2002, 2001, Fiksdal et al. 1994b, George et al. 2001). In contrast, the dominant sources of fecal contamination in the catchments studied herein are considered recent and aged swine manure application (MW) and ruminant feces (LKAS2) applied as non-point sources to the surface. The correlation between cultivation-based *E. coli* and GLUC values at site MW lies within those reported for sites under the impact of municipal sewage and the low association reported for karstic groundwater (Ryzinska-Paier et al. 2014). The authors assume that the runoff patterns and discharge dynamics in the studied catchment as well as the fecal contamination source type and age play a significant role. It is hypothesized that the association between the GLUC activity and culture-based *E. coli* analyses strongly depends on the observed habitat. GLUC measurement methods are capable of detecting activities from all enzymatic-active target bacteria, including the so-called VBNC (viable but non-cultivable) sub-populations, whereas culture-based methods are not (Cabral 2010). Furthermore, the highest correlations between the cultivation-based *E. coli* and GLUC values are expected in habitats under the influence of fecal pollution originating predominantly from a recent point source of contamination, such as waste-water treatment plant effluents. Additionally, the complexity of runoff patterns, as well as various sources and ages of fecal contamination, lead to varying proportions of the VBNC sub-population, which are associated with different compartments in the catchment. This will consequently weaken the associations between cultivation-based *E. coli* enumeration and enzymatic activity measurements.

4.4 GLUC as a conservative biochemical indicator of fecal pollution?

A further question focuses on the capacity of automated GLUC measurements as a conservative biochemical marker of microbial fecal pollution, targeting all types of fecal-associated cells, be they live, dead or dormant. However, there exist no data to date which enable the specific evaluation of this research question for contaminated water resources. Detailed studies need to be performed in future.

In this respect, cross-sensitivities as well as interferences of enzymatic activity by non-fecal compounds, such as algae or organic matter, have been reported previously (Biswal et al. 2003, Fiksdal and Tryland 2008, Molina-Munoz et al. 2007). These mechanisms of interference may limit the capacity of GLUC as a conservative biochemical marker for microbial fecal pollution. Togo et al. (2010) reported amplifying the inhibitory effects on GLUC activity, due to the abundance of different ions in the water samples. Furthermore, Chang et al. (1989) described the presence of fecal-derived *E. coli* that are not active with respect to beta-D-glucuronidase activity.

5. Conclusions and Perspectives

As it is currently state-of-the-art for gathering chemo-physical parameters, the automation of microbial monitoring will most likely become increasingly relevant in the near future. In this respect, the automated determination of enzymatic activities has great potential to make a significant contribution to complement on-line water quality monitoring. It has been clearly demonstrated that automated monitoring of enzymatic activities is now technically feasible. However, the type of substrates that can be used and their ability to serve as meaningful indicators will require further scientific evaluation.

Notably, the application of enzymatic activity monitoring is not limited to raw water resources, but is likely to generate high interest wherever enzymatic activities can be applied as a useful indicator of microbial activities. This is likely to be the case for process monitoring, such as screening for bacterial re-growth potential in engineered systems.

Further development of available substrates and the application of different substrates may also enable a more specific and diverse assessment of enzymatic activities. For example, xylanase or invertase can be evaluated as enzymes involved in C transformation, urease or amidase can be evaluated as enzymes involved in N transformation, arylsulfatase can be used as an enzyme involved in organic S transformation and alkaline phosphatase and acid phosphatase can be evaluated as enzymes involved in organic P transformation (Burns and Dick 2002, Hoppe 1991, 1983).

Furthermore, technological advances in the automated monitoring of enzymatic activity will permit a transition from static on-site operation at monitoring stations to mobile applications using portable devices.

Acknowledgements

This study was financially supported by the Austrian Science Fund (Vienna Doctoral Programme on Water Resource Systems, W 1219-N22 and project P23900-B22), the Vienna University of Technology (innovative project), the Austrian Research Promotion Agency, the research project 'Groundwater Resource Systems Vienna' in cooperation with the Vienna Waterworks, and the FP7 KBBE EU project (AQUAVALENS).

We thank Alexander Haider (xamgacom) for his technical support regarding BACTcontrol, as well as Monika Kumpan, Günther Schmid, Silvia Jungwirth (Institute for Land and Water Management Research, Federal Agency for Water Management), Markus Oismüller (Center for Water Resource Systems) and Lukas Nemeth (TU Wien) for their assistance in Petzenkirchen.

This is a joint investigation of the Interuniversity Cooperation Center for Water & Health (http://www.waterandhealth.at).

Keywords: Enzymatic activities, automatization, beta-D-glucuronidase, *E. coli*, fecal pollution, health-related water quality

References

Bergmeyer, H.-Ui. 2012. Methods of Enzymatic Analysis. Elsevier.

Biswal, N., Gupta, S., Ghosh, N. and Pradhan, A. 2003. Recovery of turbidity-free fluorescence from measured fluorescence: An experimental approach. Opt. Express 11: 3320–3331. doi:10.1364/OE.11.003320.

Blöschl, G., Blaschke, A.P., Broer, M., Bucher, C., Carr, G., Chen, X., Eder, A., Exner-Kittridge, M., Farnleitner, A., Flores-Orozco, A., Haas, P., Hogan, P., Kazemi Amiri, A., Oismüller, M., Parajka, J., Silasari, R., Stadler, P., Strauß, P., Vreugdenhil, M., Wagner, W. and Zessner, M. 2015. The Hydrological Open Air Laboratory (HOAL) in Petzenkirchen: A hypotheses driven observatory. Hydrol. Earth Syst. Sci. Discuss 12: 6683–6753. doi:10.5194/hessd-12-6683-2015.

Burns, R.G. and Dick, R.P. 2002. Enzymes in the Environment: Activity, Ecology and Applications. CRC Press.

Cabral, J.P.S. 2010. Water microbiology. Bacterial pathogens and water. Int. J. Environ. Res. Public Health 7: 3657–3703.

Chang, G.W., Brill, J. and Lum, R. 1989. Proportion of beta-D-glucuronidase-negative *Escherichia coli* in human fecal samples. Appl. Environ. Microbiol. 55: 335–339.

Exner-Kittridge, M., Salinas, J.L. and Zessner, M. 2013. An evaluation of analytical stream bank flux methods and connections to end-member mixing models: A comparison of a new method and traditional methods. Hydrol. Earth Syst. Sci. Discuss 10: 10419–10459. doi:10.5194/hessd-10-10419-2013.

Farnleitner, A.H., Hocke, L., Beiwl, C., Kavka, G.C., Zechmeister, T., Kirschner, A.K.T. and Mach, R.L. 2001. Rapid enzymatic detection of *Escherichia coli* contamination in polluted river water. Lett. Appl. Microbiol. 33: 246–250. doi:10.1046/j.1472-765x.2001.00990.x.

Farnleitner, A.H., Hocke, L., Beiwl, C., Kavka, G.G. and Mach, R.L. 2002. Hydrolysis of 4-methylumbelliferyl-β-d-glucuronide in differing sample fractions of river waters and its implication for the detection of fecal pollution. Water Res. 36: 975–981. doi:10.1016/S0043-1354(01)00288-3.

Farnleitner, A.H., Ryzinska-Paier, G., Reischer, G.H., Burtscher, M.M., Knetsch, S., Rudnicki, S., Dirnböck, T., Kuschnig, G., Mach, R.L. and Sommer, R. 2010. *Escherichia coli* and enterococci are sensitive and reliable indicators for human, livestock and wildlife faecal pollution in alpine mountainous water resources. Journal of Applied Microbiology 109: 1599–1608.

Fiksdal, L., Pommepuy, M., Caprais, M.P. and Midttun, I. 1994a. Monitoring of fecal pollution in coastal waters by use of rapid enzymatic techniques. Appl. Environ. Microbiol. 60: 1581–1584.

Fiksdal, L., Pommepuy, M., Caprais, M.P. and Midttun, I. 1994b. Monitoring of fecal pollution in coastal waters by use of rapid enzymatic techniques. Appl. Environ. Microbiol. 60: 1581–1584.

Fiksdal, L. and Tryland, I. 2008. Application of rapid enzyme assay techniques for monitoring of microbial water quality. Curr. Opin. Biotechnol. 19: 289–294. doi:10.1016/j.copbio.2008.03.004.

Fishman, W.H. and Bergmeyer, H.U. 1974. B-glucuronidase. Methods Enzym. Anal. 2: 929.

Garcia-Armisen, T., Lebaron, P. and Servais, P. 2005. Beta-D-glucuronidase activity assay to assess viable *Escherichia coli* abundance in freshwaters. Lett. Appl. Microbiol. 40: 278–282. doi:10.1111/j.1472-765X.2005.01670.x.

George, I., Petit, M. and Servais, P. 2000. Use of enzymatic methods for rapid enumeration of coliforms in freshwaters. J. Appl. Microbiol. 88: 404–413. doi:10.1046/j.1365-2672.2000.00977.x.

George, I., Crop, P. and Servais, P. 2001. Use of beta-D-galactosidase and beta-D-glucuronidase activities for quantitative detection of total and fecal coliforms in wastewater. Can. J. Microbiol. 47: 670–675.

Hoppe, H.-G. 1983. Significance of exoenzymatic activities in the ecology of brackish water: Measurements by means of methylumbelliferyl-substrates. Mar. Ecol. Prog. Ser. 11: 299–308. doi:10.3354/meps011299.

Hoppe, H.-G. 1991. Microbial extracellular enzyme activity: A new key parameter in aquatic ecology. pp. 60–83. *In*: Chróst, R.J. (ed.). Microbial Enzymes in Aquatic Environments, Brock/Springer Series in Contemporary Bioscience. Springer New York.

Kirschner, A.K.T., Kavka, G., Reischer, G.H., Sommer, R., Blaschke, A.P., Stevenson, M., Vierheilig, J., Mach, R.L. and Farnleitner, A.H. 2014. Microbiological water quality of the Danube river: Status quo and future perspectives. pp. 439–468. *In*: Liska, I. (ed.). The Danube River Basin, The Handbook of Environmental Chemistry. Springer Berlin Heidelberg.

Koschelnik, J., Vogl, W., Epp, M. and Lackner, M. 2015. Rapid Analysis of β-D-Glucuronidase Activity in Water Using Fully Automated Technology, pp. 471–481. doi:10.2495/WRM150401.

Lebaron, P., Henry, A., Lepeuple, A.-S., Pena, G. and Servais, P. 2005. An operational method for the real-time monitoring of *E. coli* numbers in bathing waters. Mar. Pollut. Bull. 50: 652–659. doi:10.1016/j.marpolbul.2005.01.016.

Molina-Munoz, M., Poyatos, J.M., Vilchez, R., Hontoria, E., Rodelas, B. and Gonzalez-Lopez, J. 2007. Effect of the concentration of suspended solids on the enzymatic activities and biodiversity of a submerged membrane bioreactor for aerobic treatment of domestic wastewater. Appl. Microbiol. Biotechnol. 73: 1441–1451.

Ouattara, N.K., Passerat, J. and Servais, P. 2011. Faecal contamination of water and sediment in the rivers of the Scheldt drainage network. Environ. Monit. Assess. 183: 243–257. doi:10.1007/s10661-011-1918-9.

Reischer, G.H., Haider, J.M., Sommer, R., Stadler, H., Keiblinger, K.M., Hornek, R., Zerobin, W., Mach, R.L. and Farnleitner, A.H. 2008. Quantitative microbial faecal source tracking with sampling guided by hydrological catchment dynamics. Environ. Microbiol. 10: 2598–2608. doi:10.1111/j.1462-2920.2008.01682.x.

Rompré, A., Servais, P., Baudart, J., de-Roubin, M.R. and Laurent, P. 2002. Detection and enumeration of coliforms in drinking water: current methods and emerging approaches. J. Microbiol. Methods 49: 31–54.

Ryzinska-Paier, G., Lendenfeld, T., Correa, K., Stadler, P., Blaschke, A.P., Mach, R.L., Stadler, H., Kirschner, A.K.T. and Farnleitner, A.H. 2014. A sensitive and robust method for automated on-line monitoring of enzymatic activities in water and water resources. Water Sci. Technol. J. Int. Assoc. Water Pollut. Res. 69: 1349–1358. doi:10.2166/wst.2014.032.

Stadler, H., Klock, E., Skritek, P., Mach, R.L., Zerobin, W. and Farnleitner, A.H. 2010. The spectral absorption coefficient at 254 nm as a real-time early warning proxy for detecting faecal pollution events at Alpine Karst water resources. Water Sci. Technol. J. Int. Assoc. Water Pollut. Res. 62: 1898.

Stadler, H., Skritek, P., Sommer, R., Mach, R.L., Zerobin, W. and Farnleitner, A.H. 2008. Microbiological monitoring and automated event sampling at Karst Springs using LEO-satellites. Water Sci. Technol. J. Int. Assoc. Water Pollut. Res. 58: 899.

Stadler, P., Blöschl, G., Vogl, W., Koschelnik, J., Epp, M., Lackner, M., Oismüller, M., Kumpan, M., Nemeth, L., Strauss, P., Sommer, R., Ryzinska-Paier, G., Farnleitner, A.H. and Zessner, M. 2016. Real-time monitoring of beta-D-glucuronidase activity in sediment laden streams: A comparison of prototypes. Water Research 101: 252–261. doi:10.1016/j.watres.2016.05.072.

Togo, C.A., Wutor, V.C. and Pletschke, B.I. 2010. Properties of *in situ Escherichia coli*-D-glucuronidase (GUS): Evaluation of chemical interference on the direct enzyme assay for faecal pollution detection in water. Afr. J. Biotechnol. 5.

Vierheilig, J., Frick, C., Mayer, R.E., Kirschner, A.K.T., Reischer, G.H., Derx, J., Mach, R.L., Sommer, R. and Farnleitner, A.H. 2013. Clostridium perfringens is not suitable for the indication of fecal pollution from ruminant wildlife but is associated with excreta from non-herbivorous animals and human sewage. Appl. Environ. Microbiol. 79: 5089–5092. doi:10.1128/AEM.01396-13.

Zibuschka, F., Lendenfeld, T. and Lindner, G. 2010. Near-Real-Time Monitoring von *E. coli* in Wasser. Österr. Wasser-Abfallwirtsch 62: 215–219. doi:10.1007/s00506-010-0240-z.

3

Advances in Electrochemically Active Bacteria: Physiology and Ecology

A.C. Marques,[1,2] *L. Santos,*[1] *J.M. Dantas,*[2] *A. Gonçalves,*[1] *S. Casaleiro,*[1,2] *R. Martins,*[1] *C.A. Salgueiro*[2,*] and *E. Fortunato*[1,*]

1. Introduction

The discovery of microorganisms with the ability of Extracellular Electron Transfer (EET), nearly three decades ago (Arnold et al. 1988, Lovley and Phillips 1988, Myers and Nealson 1988), sparked interest due to their ability to be used in diverse applications that can range from bioremediation to electricity production in Microbial Fuel Cells (MFC) (Erable et al. 2010).

These microorganisms—found in oceans, lakes, river sediments and domestic and industrial wastewater streams (Ringeisen et al. 2010)—are

[1] i3N/CENIMAT, Department of Materials Science, Faculty of Sciences and Technology, Universidade NOVA de Lisboa and CEMOP/UNINOVA, Campus de Caparica, 2829-516 Caparica, Portugal.
E-mail: emf@fct.unl.pt
[2] UCIBIO-REQUIMTE, Department of Chemistry, Faculty of Sciences and Technology, Universidade NOVA de Lisboa, Campus de Caparica, 2829-516 Caparica, Portugal.
E-mail: csalgueiro@fct.unl.pt
* Corresponding authors

known as Electrochemically Active Bacteria (EAB)[1]. Typically, the final electron acceptors are soluble compounds, such as fumarate, nitrate, sulfate and oxygen (Regan and Logan 2006, Babauta et al. 2012). However, when soluble electron acceptors are unavailable, these bacteria use the insoluble compounds as terminal electron acceptors. In this situation, these microorganisms send their electrons towards the cell exterior to reduce, for example, metals such as iron and manganese oxides (Reguera et al. 2005, Bretschger et al. 2007, Fennessey 2010, Babauta et al. 2012). Another interesting group of extracellular electron acceptors that can be used by some EAB includes electrode surfaces poised at oxidizing redox potentials (Caccavo et al. 1994).

The electrogenic properties and some aspects of EET have already been defined for pure cultures of organisms, such as *Geobacter sulfurreducens*, *Shewanella putrefaciens*, *Rhodoferax ferrireducens*, *Rhodopseudomonas palustris* DX-1 and *Ochrobactrum anthropic* YZ-1. The current list of confirmed EAB includes representatives of four of the five classes of Proteobacteria, except the ε-proteobacteria, as well as representatives of the Firmicutes and Acidobacteria (Fedorovich et al. 2009).

Although an increasing number of EAB have already been identified, isolated and characterized, this number is still very limited regarding their ubiquity in the environment. This fact has significantly constrained the fundamental knowledge about these bacteria and their role in the environment.

2. Electrochemically Active Bacteria

Microbial respiration is based on electron transfer from a donor to an electron acceptor, through a series of stepwise electron transfer events that generate the necessary metabolic energy. The donors and acceptors of electrons are typically soluble substances. However, EAB can use insoluble electron acceptors, such as minerals and metals (Babauta et al. 2012).

2.1 Extracellular electron acceptors

2.1.1 Naturally occurring insoluble electron acceptors

Iron and manganese are poorly soluble minerals that are abundant in the biosphere. These minerals can be used by EAB as terminal electron acceptors

[1] EAB are also mentioned in the literature under several other designations, such as electricigens, electrochemically active microbes, exoelectrogens and anode-respiring or anodophilic species.

for growth (Myers and Nealson 1988, Lovley et al. 2004, Stams et al. 2006). The reduction of iron and manganese minerals plays an important role in biogeochemical cycles and in bioremediation of polluted soils.

Several Fe(III)- and Mn(IV)-reducing microorganisms have already been described (Lovley et al. 2004): *Geobacter* and *Shewanella* are well-known genera with the ability to respire with minerals. Also members of other physiological groups can reduce iron, e.g., microorganisms that are known for their ability to grow by dehalorespiration, sulfate reduction (Holmes et al. 2004) and methane formation (Bond and Lovley 2003, Stams et al. 2006).

2.1.2 Other solid surfaces

EET also plays an important role in the anaerobic corrosion of steel where Fe(0) acts as electron donor for different types of respiration, including sulfate reduction and methanogenesis (Stams et al. 2006). Anaerobic corrosion of steel is caused either indirectly by an hydrogen sulfide attack yielding hydrogen and iron sulfide or directly by bacterial hydrogen consumption affecting the equilibrium of the chemical reaction: $Fe(0) + 2H^+ \rightarrow Fe(II) + H_2$ (Stams et al. 2006). Additionally, evidence for the hypothesis that sulfate-reducing bacteria obtain electrons directly from metallic iron rather than via H_2 were also presented (Dinh et al. 2004). Anaerobic corrosion is an important environmental problem, with very high costs associated in off-shore industrial processes, for example.

EET is also important in the reduction of soluble electron acceptors that form insoluble compounds upon reduction. Examples are the reduction of selenate to Se(0) and the reduction of arsenate to orpiment (As_2S_3) and realgar (AsS) (Stolz and Oremland 1999), while studying metal reduction by *Sulfurospirillum* precipitates outside the cell were observed during selenite and arsenate reduction (Luijten et al. 2004). The electrons need to be transported outside the cell in order to avoid precipitation of the reduced compounds in the cell.

2.1.3 Electrodes

Chemically rather inert graphite electrodes may both donate and accept electrons from microorganisms (Stams et al. 2006). The observation that bacteria can grow by the oxidation of organic compounds (e.g., glucose, acetate) by using electrodes as terminal electron acceptor is an intriguing example of EET, which resembles electron transfer to insoluble electron donors and electron-accepting microorganisms (Bond and Lovley 2003, Lovley 2006a).

Some microbial communities that were enriched in MFC have already been characterized (Holmes et al. 2004, Rabaey et al. 2004). The types of bacteria that are enriched in microbial fuel cells are strongly dependent on the substrate and environmental conditions. For example, if electrodes are placed in marine sediments, the *Desulfuromonas* species predominate, whereas if the electrodes are placed in freshwater sediments, the *Geobacter* species predominate. Although *Geobacter* and *Desulfuromonas* species have similar physiologies, *Desulfuromonas* prefer marine salinity, while *Geobacter* favor freshwater (Lovley 2006b, Stams et al. 2006).

2.2 Extracellular electron transfer mechanisms

The EET mechanism between bacteria and terminal acceptors located outside the cells is not unique for all EAB described so far (for a review see Hernandez and Newman 2001). Several mechanisms have been proposed, as they involve a complex set of parameters, ranging from environmental chemistry to species-specific electron transfer strategies.

A general scheme (Fig. 1) has been proposed involving electron transfer between bacteria and the solid material. This transfer is (i) directly through the components of the cell membrane, or (ii) indirectly through electrochemical mediators, secreted by the bacteria or already present in the environment (Erable et al. 2010).

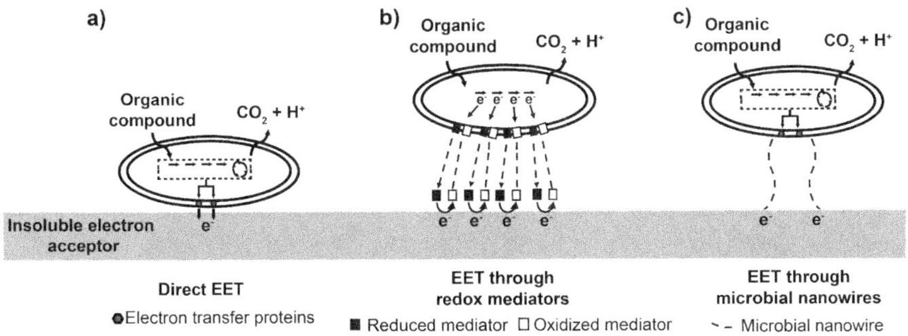

Figure 1. Mechanisms for EET: (a) direct EET; (b) indirect EET through redox mediators; (c) indirect EET through microbial nanowires. Adapted from Erable and co-workers (Erable et al. 2010).

2.2.1 Direct EET

The first observations that bacteria were able to adhere to iron mineral surfaces provided evidence that electron transfer proteins might be involved in the direct transfer of electrons to those terminal acceptors (Fig. 1a) (Stams et al. 2006).

In fact, some bacteria have already been described as being able to transfer electrons directly to electrode surfaces, e.g., *Desulfuromonas acetoxidans* (Bond et al. 2002), *Geobacter sulfurreducens*, *Geobacter metallireducens*, *Rhodoferax ferrireducens*, *Desulfobulbus propionicus* (Holmes et al. 2004), *Enterococcus gallinarum* (Kim et al. 2005).

Geobacteraceae are among the few well-studied, pure-culture, metal-reducing that are able to transfer electrons to electrode surfaces (Bond et al. 2002, Bond and Lovley 2003, Lovley 2008). Most of what is mechanistically known about *Geobacter,* EET processes is derived from pure-culture studies with the model representative, *G. sulfurreducens.* This was the first *Geobacter* species for which methods for genetic manipulation were developed, making it the choice for functional genomic studies designed to understand *Geobacter* metabolism, gene regulation and EET (Lovley et al. 2011). Gene knock-out studies on strains of *G. sulfurreducens* carrying deletions in genes encoding outer membrane *c*-type cytochromes (e.g., OmcB, OmcS and OmcE), known to be required for optimal electron transfer to metal oxides, did not show a direct correspondence with current production (Borole et al. 2011). Cytochome OmcS was proposed to be involved in a direct contact mechanism based on the partial defect in current production of an OmcS-OmcT double mutant. Cell monolayers of *G. sulfurreducens* were immobilized on electrode surfaces and characterized electrochemically to demonstrate a direct-contact mechanism (Marsili et al. 2008, Srikanth et al. 2008). Despite repeated washes to remove the loosely-bound outer membrane *c*-type cytochromes, such as OmcS, the cells were still able to interface electronically with the electrode surface (Srikanth et al. 2008). Thus, a minimum set of redox-active molecules was present in the cells that enabled direct and immediate transfer of electrons from the cell to the electrode. The abundancy of the outer membrane *c*-type cytochromes in this microorganism suggests that the pathway may not be specific. Hence, the quantity and diversity of cytochromes, rather than the presence of specific types, may enable cells to attach to the electrode surface and strategically position the redox centers in close proximity to the electrode surface for promoting optimal tunneling rates (Borole et al. 2011).

Another metal-reducing organism, that also has been extensively used as a genetically tractable model for EAB, is the *Shewanella oneidensis* MR-1 bacterium. This bacterium can directly contact and reduce both metal oxides and electrodes, using a set of *c*-type cytochromes that electronically connect the cell's cytoplasm with the extracellular electron acceptor. It was proposed that the final step that connects the bacterium periplasm and extracellular electron acceptors in *S. oneidensis* involves the outer membrane MtrCAB-OmcA complex (Borole et al. 2011, Paquete et al. 2014). In this heterotrimeric outer membrane complex MtrA, located in the periplasm and associated with the porin MtrB, receives electrons either directly from inner membrane tetraheme cytochrome CymA via a small periplasmic tetraheme cytochrome (STC) and/or via a periplasmic flavocytochrome c_3 and facilitates the passage of electrons to MtrC at the microbe-mineral interface. MtrC can then transfer electrons to the terminal acceptor or to cytochrome OmcA, as recently suggested by Paquete and co-workers (Paquete et al. 2014).

Deletions in the periplasmic MtrA *c*-type cytochrome and the MtrB β-barrel porin abolished the reduction of both metal oxides and electrodes, consistent with an overlapping mechanism for the reduction of the insoluble electron acceptors. Interestingly, although a double MtrC-OmcA mutant had a severe defect in current production, it was also defective in attachment to the electrodes, suggesting that the defect in current production may have been indirect (Bretschger et al. 2007). In fact, a direct contact mechanism for electrode reduction by *S. oneidensis* was tested in studies with nano-electrode arrays designed to permit or prevent physical contact between the cell and the electrode (Borole et al. 2011). In these studies, current was recorded before the cell membrane established direct physical contact with the electrode surface and similar current profiles were obtained with electrodes that permitted or prevented cell-electrode contact. These results suggested that mediated mechanisms drive current production by *S. oneidensis* even when cells are in physical contact with the electrode. Electrochemical analyses of cell monolayers (Baron et al. 2009) or single cells of *S. oneidensis* immobilized on the electrode provided evidence for a direct contact mechanism. The outer membrane *c*-type cytochromes, MtrC and OmcA, were required for direct EET by cell monolayers, though evidence supported a transition from a direct contact to a mediator mechanism via secreted flavins as a function of the applied potential (Baron et al. 2009). Although these results supported a direct contact mechanism via outer membrane *c*-type cytochromes in *Shewanella*, the direct pathway was slower compared to a mediator mechanism, particularly at low redox potential values (Baron et al. 2009). The type of electrode used in these various systems is likely to have selected for the most efficient electron transfer mechanism (direct or mediated), since different materials may affect the orientations of outer membrane cytochromes heme groups and tunneling rates (Borole et al. 2011).

2.2.2 Indirect EET

Electron transfer by electrochemical mediators: The transfer of electrons between the outer membrane redox-active proteins and the terminal electron acceptor is limited by the maximum distance (14 Å) that promotes optimal tunnelling rates (Borole et al. 2011). As the biofilm grows, more electroactive cells are positioned further away from the insoluble electron acceptor, thereby limiting electron transfer unless indirect mechanisms are working (Borole et al. 2011). To overcome these limitations, electrochemical mediators are added—exogenous mediators—or naturally secreted by some bacteria—endogenous mediators—which function as electron shuttles and alleviate the need to directly contact the terminal electron acceptor (Fig. 1b) (Erable et al. 2010, Borole et al. 2011). Electrochemical mediators are molecules that can be oxidized and reduced with successive recycling. In their oxidized form, they are able to cross the cell membrane to be reduced. Once in the reduced form, the mediator is exported to cell exterior to reduce the final electron acceptor (Erable et al. 2010).

2.2.3 Exogenous mediators

Artificial electrochemical mediators were often used with bacterial species, such as *Escherichia coli*, *Pseudomonas*, *Proteus* and *Bacillus* species, which were unable to transfer electrons from their internal metabolism outside the cell (Erable et al. 2010). Furthermore, exogenous mediators can also be used to enhance the EET in EAB. The most used mediators are thionine, neutral red, 2hydroxy1,4naphthoquinone and different kinds of phenazin, e.g., melanin and riboflavin (Stams et al. 2006, Erable et al. 2010).

Moreover, it was also demonstrated that humic substances can enhance the microbial electron transfer by EAB (Coates et al. 1996). Humic substances are an ubiquitous, significant and chemically heterogenous fraction of organic compounds present in aquatic and terrestrial environments (Hernandez and Newman 2001). They contain quinone structures, which can be reduced to a corresponding hydroquinone form, which can be used as an electron donor for anaerobic microbial respiration. Humic substances can therefore mediate in the EET from a bacterium to an insoluble electron acceptor or to another bacterium (Stams et al. 2006). Sulfur compounds can also act as redox mediators, e.g., cysteine-cystine shuttle and the sulfide-sulfur (polysulfide) shuttle (Stams et al. 2006).

However, many downsides are associated with the use of some exogenous mediators. First, it has not been demonstrated that microorganisms are able to maintain their growth in the presence of electrochemical mediators. Second, biological fuel cells operate continuously, requiring the permanent presence of mediators and increasing the cost of their applications. Third, the addition of artificial mediators does not always enhance the EET to the same extent (Stams et al. 2006). Finally, some exogenous mediators can be toxic and cannot be released into the environment without treatment (Erable et al. 2010).

2.2.4 Endogenous mediators

Some microorganisms, such as *Pseudomonas* species (Rabaey et al. 2004), *Shewanella putrefaciens* (Kim et al. 1999, 2002) or *Geothrix fermentans* (Bond and Lovley 2005) are able to produce electrochemical mediators to increase the EET (Hernandez et al. 2004).

EAB strains of *Pseudomonas* have also been isolated from anode biofilms (Rabaey et al. 2004) where electron transfer is driven by a mediator mechanism via phenazines. Genetic studies suggested that pyocyanin and phenazine-1-carboxamide were the redox mediators secreted by the bacteria. Pyocyanin is known to influence the intracellular redox state of the cell and carbon flux. Thus, the positive effects of pyocyanin supplementation on current production are likely to be due, at least in part, to improved metabolic fluxes in the cell, which control carbon utilization and electron donors available for EET, rather than improved ET mechanisms (Borole et al. 2011). The bacterium *Pseudomonas aeruginosa* has been described as producing phenazin molecules, increasing electron transfer rates measured in MFCs (power obtained 116 mW/m^2). In contrast, mutant strains of *P. aeruginosa*, which could not synthesize electrochemical mediators have achieved only 6 mW/m^2 representing only 5 per cent of the power observed with the non-deficient strain (Rabaey et al. 2005, Erable et al. 2010).

Shewanella spp. is known to reduce insoluble metals at a distance and without the need to physically contact the insoluble electron acceptor (Nevin and Lovley 2002, Lies et al. 2005). *Shewanella* spp. excrete flavin mononucleotide (FMN) and, to a lesser extent, riboflavin mediators for the reduction of Fe(III) oxides (Von Canstein et al. 2008). Genetic studies demonstrated the involvement of the Mtr respiratory pathway (Fig. 2) in the reduction of both the flavin mediators and the electrode in MFCs, with MtrC playing a key role in flavin reduction and OmcA, in electrode reduction and attachment (Baron et al. 2009). Although diffusion limitations are predicted to make a mediator mechanism inefficient, the synthesis of riboflavin by *S. oneidensis* MR-1 increased the rates of electron transfer by 370 per cent while requiring just a fraction (<0.1 per cent) of the cell's ATP budget (Borole et al. 2011).

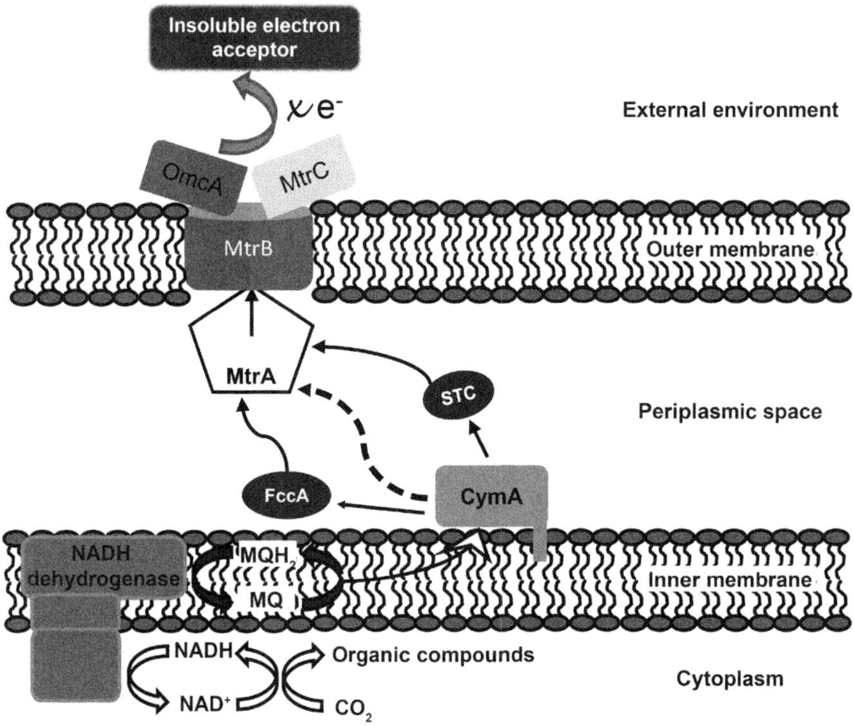

Figure 2. Electron flow in Mtr pathway. The arrows indicate the proposed flow for electrons. Solid arrows connecting periplasmic components indicate the pair wise interactions identified by NMR experiments (Fonseca et al. 2013). Adapted from Santos and co-workers (Santos et al. 2015) and Venkata Mohan and co-workers (Venkata Mohan et al. 2014).

To date, the proposed model for extracellular electron transfer in *Shewanella oneidensis* (Venkata Mohan et al. 2014, Santos et al. 2015) considers that the oxidation of organic molecules releases electrons to the menaquinone (MQ) pool via NADH dehydrogenase. From this point, a network of *c*-type cytochromes is responsible for the long-range electron transfer from the MQH_2 pool to extracellular acceptors. CymA is proposed to accept electrons from the MQH_2 pool, which are then delivered to multiheme periplasmic cytochromes, which establish the interface between the cytoplasmic and the outer membrane-associated electron transfer components. Both STC (small tetraheme cytochrome *c*) and FccA (flavocytochrome *c*) interact with their redox partners, CymA and MtrA, through a single heme (Fonseca et al. 2013).

In contrast to *Shewanella* and *Geotrix* species that produce shuttles permitting reduction iron oxides at a distance, evidence was collected of the need for direct contact between *Geobacter* species and exogenous electron acceptors. Although several cytochromes essential for EET in *Geobacter* have been identified, the electron transport from the cellular interior to the exterior remains unclear. In fact, various inner membrane associated (IM), periplasmic and outer membrane associated (OM) cytochromes involved in EET have been identified in *G. sulfurreducens* (Inoue et al. 2010, Morgado et al. 2010, Pokkuluri et al. 2011, Qian et al. 2011, Seidel et al. 2012). These included the diheme cytochrome MacA, a family of five triheme (designated, PpcA-E) and a dodecaheme (GSU 1996) periplasmic cytochromes and the outer membrane cytochromes OmcS and OmcZ. Only evidence for a direct interaction between cytochromes MacA and PpcA was obtained to date (Seidel et al. 2012). However, the high homology between MacA and other diheme cytochrome *c* peroxidases and the confirmation of its peroxidase activity (Seidel et al. 2012) raises the possibility for alternative electron entry points through the inner membrane. Recently, two inner membrane cytochromes were suggested to be involved in the reduction of high (cytochrome ImcH) and low (cytochrome CbcL) extracellular electron acceptors in *G. sulfurreducens*, which can play a similar role to that attributed to cytochrome CymA in *S. oneidensis* (Levar et al. 2014, Zacharoff et al. 2016). A possible model for EET in *G. sulfurreducens* is summarized in Fig. 3.

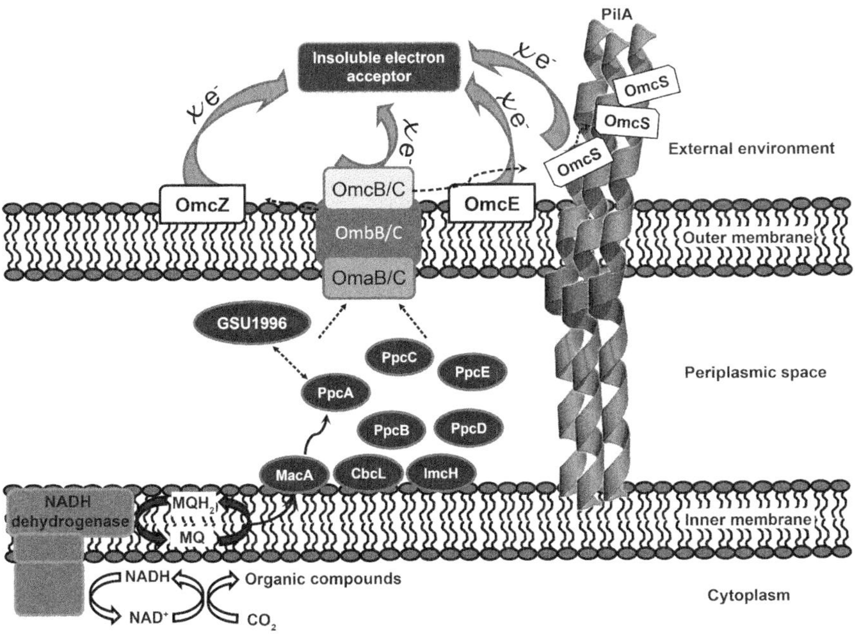

Figure 3. Proposed model for Geobacter' EET. Adapted from Santos and co-workers (Santos et al. 2015).

Electron transfer by microbial nanowires: The mechanical stability of the biofilm is provided by the biofilm matrix, a hydrated extracellular polymeric matrix that encases the biofilm cells (Flemming and Wingender 2010). The biofilm matrix could potentially offer a resistance pathway to EET unless bacteria develop strategies to increase its conductivity.

The term 'microbial nanowire' (Fig. 1c) first appeared in microbiology literature when Reguera and co-workers (Reguera et al. 2005) reported electrical conductivity measured across the diameter of a sheared pilus-like filament from *G. sulfurreducens* DL-1 that was deposited and dried onto highly ordered pyrolytic graphite (HOPG), using conducting probe atomic force microscopy (CP-AFM). The finding that *G. sulfurreducens* produced electrically conductive pili to transfer electrons to Fe(III) oxides (Reguera et al. 2005) and to form biofilms (Reguera et al. 2006) suggested that these protein appendages could mediate long-range electron transfer reactions across the anode biofilms as well. Genetic studies in *G. sulfurreducens* demonstrated that the pili were not required to attach to the anode, but were required for biofilm growth, differentiation and maximal power production (Reguera et al. 2006). Thus, the pili were proposed to establish a 'nanopower grid' that mediated long-range electron transfer through the biofilms in an energy-efficient manner. Extracellular bacterial appendages, such as pili, are important components of the biofilm matrix and contribute to its stabilization since they are also structural components of the biofilm matrix (Borole et al. 2011, Strycharz-Glaven et al. 2011).

S. oneidensis MR-1 also produces conductive appendages when grown at anode surfaces (Gorby et al. 2006). These structures conduct electrons along their length, using bound *c*-type cytochromes assembled on a yet uncharacterized filamentous scaffold (El-Naggar et al. 2010). Genetic studies support the involvement of MtrC and/or OmcA *c*-type cytochromes in the conductivity of *Shewanella's* nanowires (Gorby et al. 2006, El-Naggar et al. 2010). However, the contribution of these conductive appendages to current production in MFCs is still highly debated, based on the fact that *Shewanella* biofilms also use flavins to reduce the anode electrode—a process that depends on the activity of the MtrC and OmcA *c*-type cytochromes (Baron et al. 2009). Recent studies using nano electrode platforms and anaerobic batch cultures indicated that EET in thin anode biofilms of *S. oneidensis* MR-1 predominantly occurred via mediators (Jiang et al. 2010). Furthermore, although the biofilm cells produced filaments with a diameter such as those reported for *Geobacter's* pili (4–5 nm) (Reguera et al. 2005), they did not appear to contribute to current production (Jiang et al. 2010). By contrast, nanofilaments were abundantly produced in thick biofilms of the same strain formed on stainless steel coupons in batch aerobic cultures

(Ray et al. 2010). In as much as the production of *Shewanella* nano wires has been linked to O_2 limitation, these conductive appendages could be a specific mechanism for EET in areas with low oxygen availability (Borole et al. 2011).

There are fundamental differences but also similarities between the above-mentioned types of EET. They are similar in the sense that the electron donor is physically separated from the electron acceptor—the latter being a microorganism, an insoluble mineral or an electrode. Moreover, the differences between EET to a microorganism or to a solid electron acceptor are not as clear-cut as they seem (Stams et al. 2006). Modern molecular biological techniques shed light on the presence of phylogenetic microbial groups in anaerobic environments in which EET is a key process. However, to understand their function, researchers need to put efforts in developing novel and innovative strategies to isolate novel microorganisms to make them available for detailed physiological studies. This will improve the understanding of their role in Nature.

3.1 Applications

EAB are currently explored in many fields, including biotechnology, sustainable energy and bioremediation (Erable et al. 2010). For instance, EAB in biofilms are able to reduce heavy metals, such as chromium and uranium, play a role in bioremediation of groundwater and contaminated soils (Lovley et al. 2004, Cologgi et al. 2011).

3.1.1 Microbial fuel cells

As mentioned above, several EAB are also able to use electrodes as final electron acceptors. Therefore, the most common applications of EAB are MFC devices (Fig. 4), which generate electricity from organic compounds through microbial catabolism (Bond and Lovley 2003, Logan 2009, Erable et al. 2010, Mathuriya and Sharma 2010).

A typical MFC contains an anaerobic anode chamber and an aerobic cathode separated by a proton exchange membrane (PEM). EAB reside within the anaerobic anode chamber and the electrons generated during their respiration are delivered to the anode and then flow to the cathode through the external electrical circuit. Simultaneously, protons generated at the anode diffuse through the PEM and join the electrons released to the catholyte (e.g., oxygen, ferricyanide) at the cathode chamber (Kim et al. 2006).

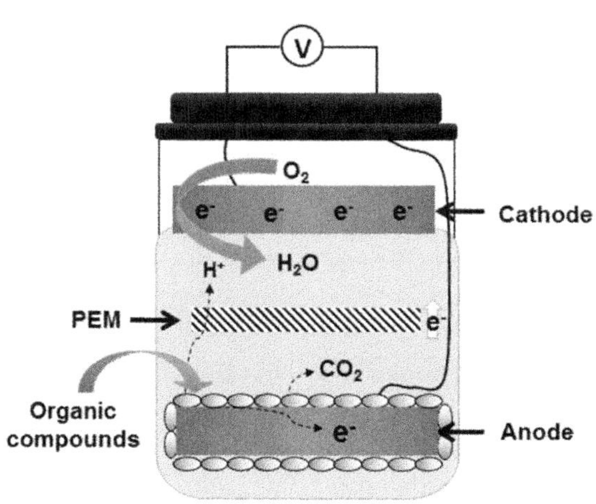

Figure 4. Schematic representation of a typical MFC.

The performance of MFC depends on several factors, such as fuel cell configuration, anolyte nature and microenvironment, electrode materials and spacing, membrane, nature of microbes, electron transfer mechanism, redox condition, pH, electron accepting condition, etc. Solution chemistry also regulates the MFC performance to a large extent (Venkata Mohan et al. 2014).

The basic setup of MFC has been configured with diverse designs to meet a wider range of applications, e.g., benthic MFC, plant MFC, stacked MFC, etc. Benthic MFCs are specifically designed to harness bioelectricity from an aquatic eco-system by using natural habitants. The fuel cell system is configured such that anode is placed in a sediment and cathode in overlying on the water layer (Venkata Mohan et al. 2014).

Plant-based MFCs facilitate indirect utilization of solar radiation to generate green electricity by integrating the roots of living plant (rhizosphere) in MFC devices. *Rhizodeposits* secreted from roots are organic in nature and provide a favorable environment for proliferation of bacteria. Plants, such as Reed mannagrass, *Pennisetum setaceum* and paddy (Kouzuma et al. 2013) have been used for electricity generation (Kouzuma et al. 2013, Venkata Mohan et al. 2014).

The electricity generated in MFC can be coupled to other applications, such as wastewater treatment process for removal of oxidizable matter from industrial and domestic wastewaters, metal extraction from minerals, autonomous sensors for long-term operations in remote locations, mobile platform robot/sensor, microscopic systems for drug-delivery, nanoparticles synthesis and seawater desalinization (Kim et al. 2003, 2007, Erable et al. 2010, Mathuriya and Sharma 2010, Khan et al. 2012).

MFC devices currently being used and studied do not generate sufficient power to support widespread and cost-effective applications (Erable et al. 2010, Borole et al. 2011). Hence, research focuses on strategies to enhance the power output of the MFC devices, including exploring more electrochemically active microbes to expand the few already known EAB families.

3.1.2 Microbial electrolysis cells

One possible solution to meet biofuel and chemical production demands is represented by the possibility of using novel technologies, such as Microbial Electrolysis Cells (MEC) (Erable et al. 2010).

Biofuels production: MEC devices emerged as an alternative route to H_2 production from renewable resources as they are a much cost-effective process yielding higher levels compared to water electrolysis (Rabaey et al. 2004). MEC differ from MFC because an external potential is applied to force electrons and protons to cross the endothermic barrier to form hydrogen gas. The protons migrate to the cathode and get reduced to form H_2 with electrons travelling from the anode under applied voltage (Babauta et al. 2012, Liu et al. 2012). Double chambered bottle-type MEC devices were designed and operated, using acetate as a substrate with H_2 yield of 2.9 mol H_2/mol acetate achieved at 0.85 V of external voltage (Liu et al. 2012). Tartakovsky and co-workers (Tartakovsky et al. 2009) achieved a fivefold increase in the H_2 production rate, using a continuous-flow membrane-less MEC.

Biocommodity products: Based on the electron accepting conditions at the cathode, different compounds can be synthesized. Therefore, MEC can also be used for product synthesis, e.g., acetate, ethanol, hydrogen peroxide, butanol, etc. (Table 1). The product formation is completely based upon the electron accepting conditions and redox potential of the system (Venkata Mohan et al. 2014). Ethanol can be formed at cathode at a redox potential of –0.28 V vs. SHE by considering acetate as electron acceptor. Likewise a diverse range of value-added products can be harnessed especially at the cathode in the absence of O_2 as electron acceptor, along with the power generation (Rabaey and Rozendal 2010). Some of the reactions require less redox potential, which can be accomplished by *in situ* generated bio-potential, while some reactions require more redox potential at the cathode. In such cases, the designated redox potential is maintained by applying small external potentials to meet the energy necessary to cross the energy barrier for product formation. Recently, several laboratory-scale studies have demonstrated the use of acetogens that have the ability to convert various syngas components (CO, CO_2 and H_2) to multi-carbon compounds,

Table 1. Bioelectrosynthesis Parameters for Some Value Added Products.

Redox System	Applied Potential (V)	Final Product
H^+/H_2	–0.41	Hydrogen
HCO_3^-/Methane	–0.24	Methane
O_2/H_2	–0.28	Hydrogen peroxide
HCO_3^-/Formate	–0.41	Formate
HCO_3^-/Acetate	–0.28	Acetate
HCO_3^-/Ethanol	–0.31	Ethanol
HCO_3^-/Butanol	–0.37	Butanol
HCO_3^-/Fumarate/Succinate	0.03	Fumarate
HCO_3^-/PHA	–0.31	PHB

such as acetate, butyrate, butanol, lactate and ethanol, in which ethanol is often produced as a minor end-product. Microaerophilic cathode in MEC devices showed enhanced electrogenesis and reduced losses due to controlled microbial metabolism which simultaneously facilitated polyhydroxyalkanoates (PHA) production (Srikanth et al. 2012).

3.1.3 Biosensors

MFC application can be extended to biosensors where the biocatalyst in the anode chamber acts as a biological detector to evaluate the performance of the system.

EAB can be useful in electrochemical biosensors to monitor the development of biofilms in remote facilities, where their presence is undesirable (Erable et al. 2010).

Advanced application in MFCs technology was reported by the development of BOD sensor (Kim et al. 2003). Chang and co-workers (Chang et al. 2005) also performed studies in MFC to be used as BOD sensor. They correlated the (Biochemical Oxygen Demand) BOD concentration with the current measured on the anode surface.

Larsen and co-workers (Larsen et al. 2000) designed and used a micro-scale biosensor for *in situ* monitoring of nitrate/nitrite concentrations. This technique was based on diffusion of nitrate/nitrite inside the concentrated mass of bacteria through a membrane. EAB converted these ions into nitrous oxide which could be electrochemically detected.

Significant optimization of the mentioned applications is still required. Further research into the physiology and ecology of EAB is essential to design microorganisms with improved electron-transfer capabilities (Lovley 2006a, Yi et al. 2009).

4. Screening Methods

Since the discovery of EET capability in some bacteria, knowledge about their ecology and physiology has been evolving over the past few decades. Consequently, the screening methods of these bacteria has also been evolving.

The first identification methods were based on the ability of these microbes to reduce iron and manganese oxides, such as plate assay-based screening techniques (Burnes et al. 1998) and 16S rRNA-targeted oligonucleotide radioactively labelled probe, specific for reduced iron and manganese (DiChristina and DeLong 1993). However, due to the different mechanisms of bacterial EET towards these compounds and carbon electrodes, such methods presented some drawbacks when used for identification of EAB. This was the case of *Pelobacter carbinolicus*, the first Fe(III) oxide-reducing microorganism found to be unable to produce current in a MFC. This observation suggested for the first time that the mechanisms for extracellular electron transfer to Fe(III) oxides and fuel cell anodes may be different (Richter et al. 2007).

Therefore, it was necessary to develop identification methods based on the electricity-producing ability of these organisms. *Clostridium butyricum* (Park et al. 2001) and *Aeromonas hydrophila* (Pham et al. 2003) were the first two microorganisms isolated from MFC anodes by plating with soluble Fe(III) citrate or Fe(III) pyrophosphate. By using insoluble Fe(III) oxide as the electron acceptor, *Geopsychrobacter electrodiphilus* was isolated from a marine sediment fuel cell (Holmes et al. 2004). Although these organisms have shown electricity generation in MFCs, Fe(III) plating methods eliminate the growth and isolation of other EAB that may not be able to respire with iron on the plates. General nutrient agar plates were also used for isolation of EAB from MFCs under aerobic and anaerobic conditions (Rabaey et al. 2004), but this method allowed non-selective growth of non-electricity producing bacteria, making it difficult to choose which colonies should be used in further studies. Therefore, the isolation methods used to obtain electricity-producing bacteria by plating are indirect and potentially biased, and they may not allow identification of the true diversity of the EAB functioning in MFCs (Zuo et al. 2008).

Recently, other screening methods based on MFC principles were developed, such as U-tube shaped MFC devices (Zuo et al. 2008), voltage-based screening assays (Biffinger et al. 2009), microfabricated MFC arrays (Hou et al. 2009, Choi et al. 2015) as well as photometric methods (Yuan et al. 2013, Marques et al. 2015).

4.1 MFC-based screenings

4.1.1 U-tube shaped MFC

A special U-tube shaped MFC (Fig. 5) was developed to enrich EAB with isolation based on dilution-to-extinct methods, which are independent of the need for metal reduction (Zuo et al. 2008). This device was constructed in a U shape in order to allow bacteria in suspension to directly settle on the anode. Instead of using two symmetrical curled tubes containing floating electrodes inside the tubes, the U-tube isolation device uses a straight anode tube with a flat anode placed at the bottom. Thus allowing a small number of cells in the most diluted samples to settle onto an anode surface so that they can grow and produce current. The asymmetric cathode chamber also provides a high ratio of cathode volume to anode volume and has a graphite fiber cathode with a large surface area to increase cathode efficiency, and a chemical catholyte solution is used to avoid oxygen contamination. Through repeated dilution to extinction, Zuo and co-workers (Zuo et al. 2008) showed that this device could be used to directly isolate EAB according to their electricity-generating ability and successfully obtained a pure culture identified as *Ochrobactrum anthropic* YZ-1, by sequence and phylogenic analysis.

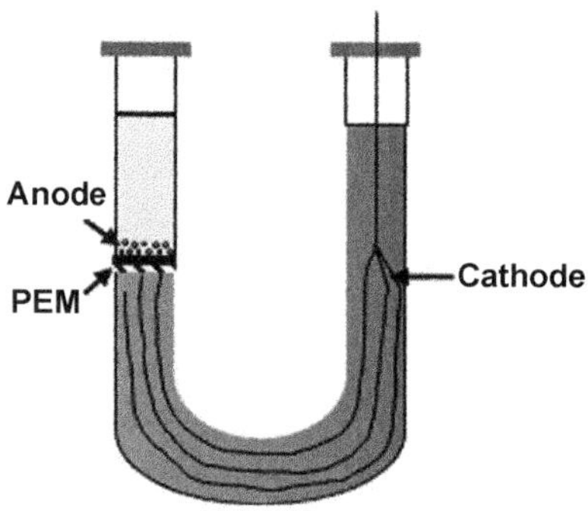

Figure 5. Schematic representation of a U-tube shaped MFC, designed by Zuo and co-workers (Zuo et al. 2008), for EAB screening.

4.1.2 Voltage-based screening assay

Biffinger and co-workers (Biffinger et al. 2009) described a four- to nine-well prototype high throughput voltage-based screening assay (VBSA) (Fig. 6), designed using MFC engineering principles and a universal cathode. This device is an operational prototype of a high-throughput screening assay that uses real-time voltage detection instead of metal reduction for potential microbial power output from MFCs. *S. oneidensis* strains MR-1 and DSP10 were used for bacterial growth studies as well as power output from various carbon fuels. The data collected resulted in efficient determination of energy-harvesting potential as compared to using large-scale individual MFCs and enabled multiple nutrients to be screened simultaneously for current and power output from *Shewanella*. The use of a universal cathode and a well-defined catholyte (potassium ferricyanide) allows small changes in voltage (and current) to be analyzed between each chamber with excellent reproducibility.

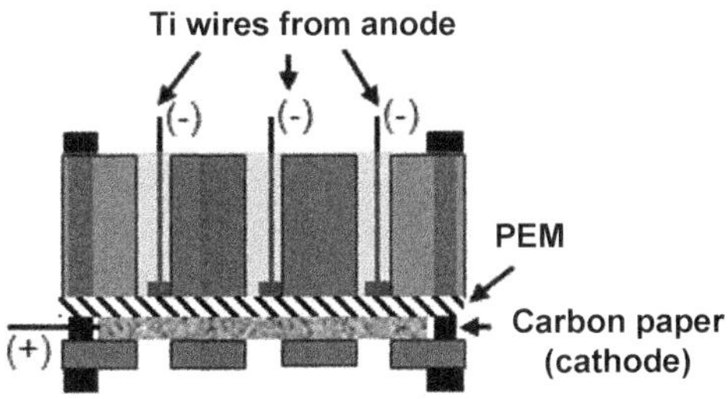

Figure 6. Schematic representation of a four-well prototype, designed by Biffinger and co-workers (Biffinger et al. 2009), for EAB screening.

4.1.3 Microfabricated MFC arrays

Hou and co-workers (Hou et al. 2009) described the development of a microfabricated MFC array (Fig. 7), a compact and user-friendly platform for the identification and characterization of EAB. The MFC array consists of 24 integrated anode and cathode chambers, which function as 24 independent miniature MFCs and support direct and parallel comparisons of microbial electrochemical activities. The electricity generation profiles of spatially distinct MFC chambers on the array loaded with *S. oneidensis* MR-1 differed by less than 8 per cent. A screen of environmental microbes using the array identified an isolate that was related to *Shewanella putrefaciens* IR-1 and *Shewanella* sp. MR-7, and displayed 2.3-fold higher power output than the *S. oneidensis* MR1 reference strain. However, this device contains

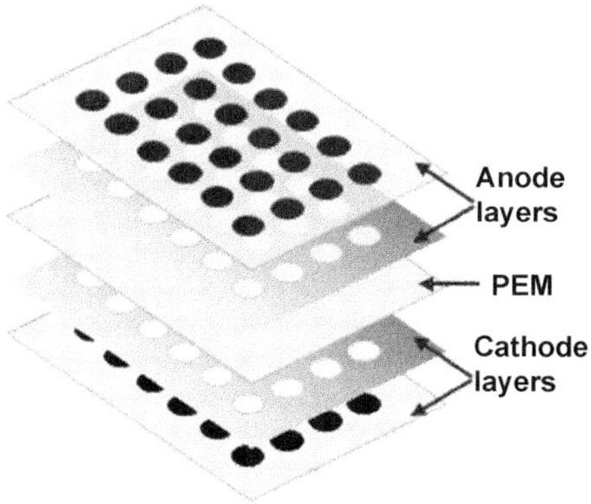

Figure 7. Schematic representation of a microfabricated MFC array, designed by Hou and co-workers (Hou et al. 2009), for EAB screening.

chambers with confined volumes (less than 1 ml) that prevented studies beyond a few days of operation due to depletion of electron acceptors and organic carbon substrates from the small chambers. So, three years later, the same author reported the first microfluidic MFC array that supports long-term operation and analysis by replenishing anolytes and catholytes in each of the reaction chambers. The microfabricated 24 chamber system utilizes integrated microfluidic channels and microvalves to periodically or continuously replenish the anolyte or catholyte of the integrated miniature MFC units. The utility of the system was validated by successfully screening microbial consortia collected from geographically diverse environments for communities that support enhanced MFC performance (Hou et al. 2012).

Even though MFC screening-based methods are very efficient for EAB screening, they have the disadvantage of being relatively slow and require expensive materials. Hence, rapid, inexpensive and easy-to-operate screening methods are required to identify EAB and to evaluate their extracellular electron transfer abilities.

5.1 Photometric method using WO_3 nanoprobes

5.1.1 Conventional 96-well plate platform

Yuan and co-workers (Yuan et al. 2013) described, for the first time, a rapid EAB identification method based on visual probe-detection technology and using an electrochromic (EC) material – tungsten trioxide (WO_3)—a deeply studied inorganic material (Lee et al. 2006, Deb 2008, Santos et al. 2014).

EC materials have the capability to change their optical properties when a sufficient electrochemical potential is applied (Bange 1999). The coloration process on WO_3 occurs by double injection of ions and electrons inside the material, as depicted in Equation 1:

$$WO_3\ (white/yellow) + xM^+ + xe^- \leftrightarrow M_xWO_3\ (blue) \qquad (1)$$

Typical cations are H^+, Li^+ and Na^+, with a stoichiometry that can vary between zero and one (Zheng et al. 2011).

The method uses the electron-transfer process of EAB to hexagonal WO_3 nanoclusters in an aqueous dispersion, assembled in a 96-well plate, where obvious coloration occurred when EAB were injected in the wells. The unique microstructure of the WO_3 nanoclusters enables a sensitive bioelectrochromic response to the number of electrons transferred. The color intensity of M_xWO_3 increases with an increase in x, which is determined by the number of electrons transferred from EAB in the bioelectrochromic process.

S. oneidensis MR-1 and ten of its mutants and *G. sulfurreducens* DL-1, typical EAB, were used to validate the proposed method. A high correlation between color intensity and electrochemical activity of bacteria was observed, with 0.833 and 0.994 in Spearman's ρ ($P<0.01$) for the correlations between color intensities and achieved current densities by different strains and the population density of the cells, respectively. These results demonstrate the effectiveness and high sensitivity of the high-throughput photometric probe method for evaluating extracellular electron transfer abilities of EAB.

Additionally, *P. carbinolicus* was also tested and no obvious difference was observed for the vials with and without bacterial incubation. Thus, unlike conventional methods based on use of metal oxides as electron acceptors that usually give false positive result, this method can effectively distinguish electricity-production from non-electricity-production bacteria and has high specificity and reliability.

Furthermore, two new EAB were successfully isolated from mixed cultures, with the described screening method: *Kluyvera cryocrescens* and *Lysinibacillus sphaericus* D-8 (He et al. 2014).

The published work used WO_3 as an effective probe for electron extraction and electron transfer-ability evaluation of EAB, thus providing a whole new method for EAB identification, which will substantially increase our capacity to identify EAB from Nature and enhance our fundamental understanding about them.

The method has several advantages, such as being rapid, having high throughput and being very effective and reliable without false positive results. Despite that, the use of a conventional microplate platform has some downsides owing to the difficulty in recording, by scan or photograph, and to analyze the results due to the possible presence of reflection and/or glares and low color contrast between the colorimetric result and the testing platform.

5.1.2 Office paper platform

Marques and co-workers (Marques et al. 2015) adapted the well-established colorimetric method to a conventional non-treated office paper based platform. Different nanostructures of WO_3, synthesized though a microwave-assisted hydrothermal method (Santos et al. 2016) were integrated in a wax-printed office paper platform as an active layer for EAB detection. As a proof of concept, these sensors were used to visually detect the presence of a well-known EAB species (Lovley et al. 2011)—*G. sulfurreducens* DL-1 (Fig. 8).

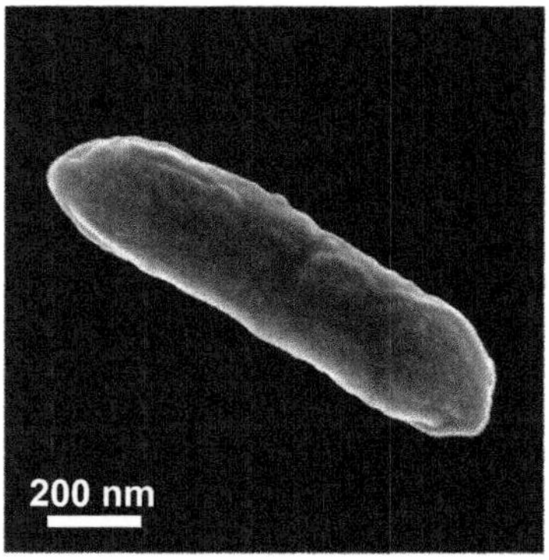

Figure 8. SEM image of *Geobacter sulfurreducens* DL-1 bacterium.

The work reports a colorimetric sensor based on Lab-on-Paper technology (Veigas et al. 2012, Costa et al. 2014). Lab-on-Paper technology was first introduced in 2007 by George Whitesides group as a method for patterning paper to create well-defined, millimeter-sized channels, comprising hydrophilic paper bound by hydrophobic polymer, photoresist or wax (Martinez et al. 2007, Costa et al. 2014). In this paper-based platform, wax printing was used to define reaction zones for EAB identification, since it is the fastest and simplest fabrication method reported to date. This approach uses a solid ink printer, in which the ink is supplied as solid wax that is melted before being ejected from the print head and solidifies immediately onto the paper surface. The technology generates up to 90 per cent less printing waste than comparable color laser printers because there are no cartridges to dispose of and less packaging to add to landfills. Solid ink is formulated from a non-toxic resin-based polymer and is safe to handle. The printed paper is then processed on a hot plate (140°C, 2 min), allowing the wax to diffuse vertically through all the paper thickness, creating hydrophobic barriers that define hydrophilic reaction zones (Martinez et al. 2010, Jokerst et al. 2012, Liana et al. 2012, Costa et al. 2014). Lastly, a WO_3 NPs aqueous dispersion is drop casted in the defined reaction zones, ensuring no cross-contamination between adjacent samples as well as confinement of the WO_3 nanoparticles dispersion to one particular area (Fig. 9a).

The EAB screening in the paper-based platform is a very simple process and is schematized in Fig. 9b.

The use of paper as a platform for biological assays brings numerous advantages, such as being in expensive, eco-friendly, easy to operate and analyze, disposable and easy to transport as compared to plastic and glass substrates (Liana et al. 2012, Zhou et al. 2014). To date, researchers have focused on developing paper-based sensors with less complicated fabrication techniques and operation such that it can be used in developing world applications where simple and easy to operate devices are highly desirable (Martinez et al. 2007, 2010, Pelton 2009, Jokerst et al. 2012, Veigas et al. 2012, Zhang and Rochefort 2013, Costa et al. 2014).

Marques and co-workers showed that office paper allows a superficial adhesion of the WO_3 nanoparticles, which facilitate the interaction of EAB with electrochromic nanoparticles, promoting an intense and uniform coloration on the reaction zone, contrary to chromatography paper. Office paper is optimized for printing and therefore, has a more uniform surface, lower porosity and higher hydrophobicity (water contact angle of 106°) when compared to chromatography paper (water contact angle of 12°), the most common type employed in paper-based devices. Figure 10 shows the office and chromatography paper morphologies without (Fig. 10a and c) and with (Fig. 10b and d) drop-casted WO_3 nanoparticles. The chemical analysis for both drop-casted papers (Fig. 10c and f) clearly shows a better distribution of tungsten chemical element in the office paper surface.

The proposed paper-based sensor was successfully used to test the presence of electrochemically active *G. sulfurreducens* DL-1 (Lovley et al. 2011) cells (Fig. 9b). *G. sulfurreducens* DL-1 was chosen as a model microorganism for testing the developed paper-based sensor, since it is a well-known and

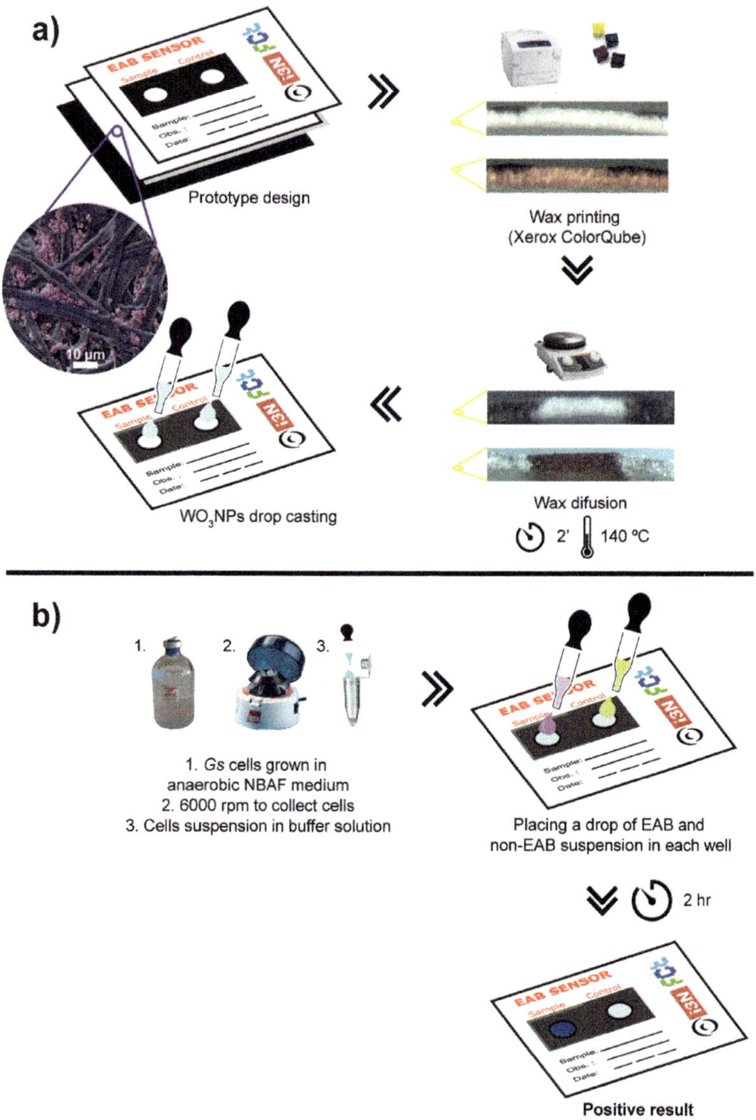

Figure 9. (a) Schematic representation of the paper-based sensor production process; (b) Schematic representation of the realization of an EAB screening.

deeply studied EAB. It can be found in distinct environments and has been widely used in the development of EAB-based applications. The presence of these cells in the sample induces an electrochromic response in WO_3 nanoparticles, characterized by a tungsten bronze formation (Equation 1), displaying a deep blue color that highly contrasts with the white background provided by the paper platform.

Figure 10. SEM images and EDS maps of the paper surfaces: (a) Office paper; (b) Office paper with WO_3 nanoparticles; (c) Tungsten element EDS map on office paper with WO_3 nanoparticles; (d) Whatman™ (cellulose filter) paper; (e) Whatman™ paper with WO_3 nanoparticles; (f) Tungsten element EDS map on Whatman™ paper with WO_3 nanoparticles.

A screening colorimetric assay with different WO_3 nanostructures (monoclinic, orthorhombic and hexagonal crystallographic structures) was performed both in the developed paper-based sensor and in the conventional 96-well plate. Additionally to Yuan and co-workers (Yuan et al. 2013) results, the electrochromic response of WO_3 to EAB cells in the conventional assay was also observed with monoclinic and orthorhombic crystallographic WO_3 structures, due to the higher concentration of nanoparticles and enhanced interaction with EAB (Marques et al. 2015).

In the paper-based platform, the electrochromic response was only achieved with hexagonal WO_3 nanoparticles (h-WO_3). The h-WO_3 nanoparticles have attracted much attention due to their well-known tunnel structure where the openness degree is higher when compared to the layered structure of orthorhombic or monoclinic geometries. This feature results in an easier intercalation of cations to form tungsten bronzes and concomitant enhancement of the electrochromic properties (Ha et al. 2009, Kharade et al. 2012). Moreover, the one-dimensional nanowire shape originates a structure with a high surface area and increased surface atom density that can easily interact with the EAB (Fig. 11).

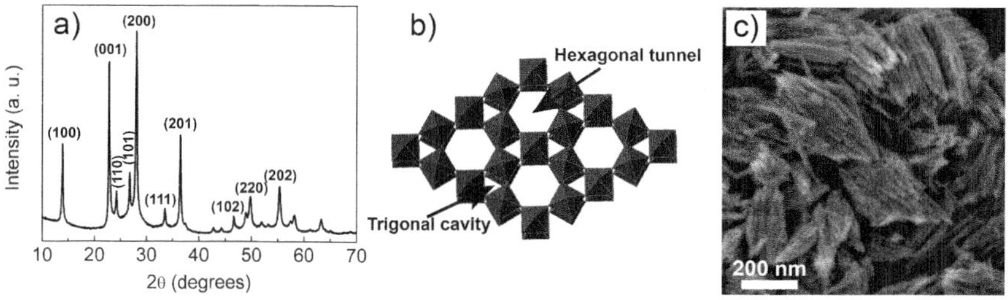

Figure 11. Hydrothermal microwave-assisted synthesized hexagonal WO_3 nanoparticles characterization: (a) X-ray diffractogram; (b) Crystalline structure produced with Crystal Maker software (Centre for Innovation & Enterprise, Oxford); (c) SEM image of WO_3 nanoclusters.

The influence of the hexagonal WO_3 nanoparticle concentration in EAB screening was evaluated (Fig. 12a). Figure 12b represents the RGB analysis of the resulting color of the *h*-WO_3 nanoparticles in contact with *G. sulfurreducens* DL-1 cells. From this analysis, it is possible to conclude that 15 g/L and 20 g/L *h*-WO_3 nanoparticles dispersion renders higher RGB ratios when compared to the other studied concentrations. Moreover, 15 g/L nanoparticle dispersion presents a linear response to the increasing *G. sulfurreducens* cells concentration.

The sensor layout can be designed to resemble a conventional microplate for parallel assays (Fig. 12a), with the same well dimensions as a 384-well plate, or prototyped into single-use sensors (Fig. 12c) containing only a test and control wells, also with the same well dimensions. Figure 12d shows the RGB analysis of a positive control (*G. sulfurreducens* DL-1), a negative control (*E. coli*) and a blank well (no sample) for background information. *G. sulfurreducens* DL-1 cells in a late-exponential phase of growth ($Abs_{600\ nm} \sim 0.5$) display an RGB ratio of 1.33 ± 0.005, while the negative control, *E. coli* under the same conditions, and blank sample display an equal ratio of $0.99 \pm 0.001/0.004$. This result reveals a clear statistically significant difference between a positive and negative sample ($P < 0.0001$), proving the specificity of the developed paper-based device (Table 2).

Moreover, it was also possible to detect EAB at latent phase ($Abs_{600\ nm} \sim 0.1$) with an RGB ratio of 1.10 ± 0.040, thus confirming that the sensor here described is sensitive to low concentrations of EAB.

Preliminary results (unpublished) also showed that this paper-based platform can be used to evaluate extracellular electron transfer capabilities. Figure 12e and f show colorimetric data obtained for three different bacteria: *G. sulfurreducens* KN400, a strain known to have improved capacity for extracellular electron transfer and current production (Yi et al. 2009, Butler et

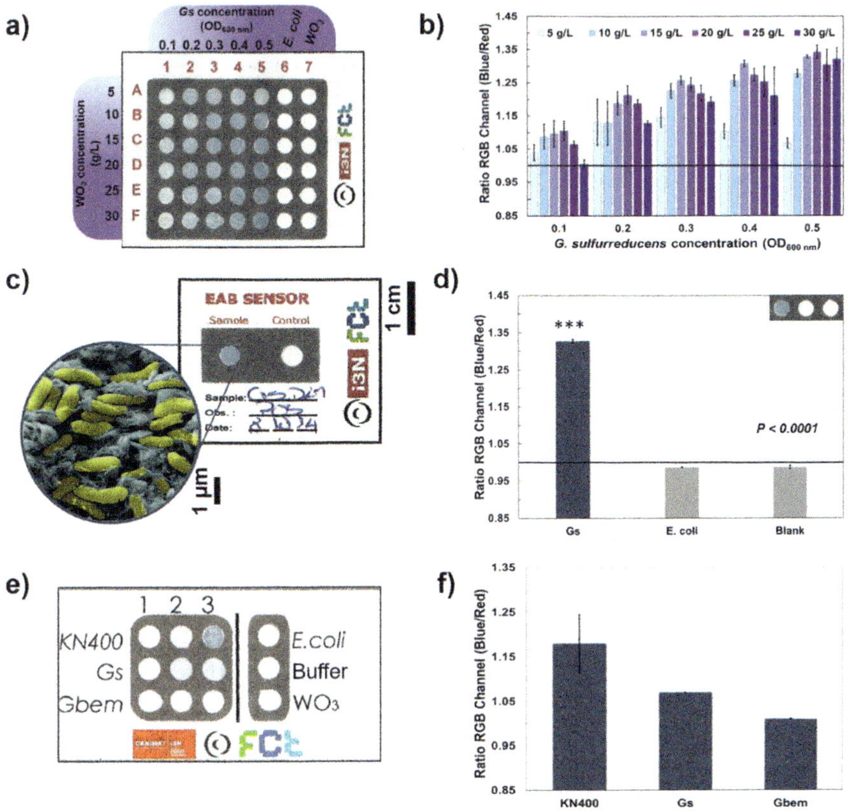

Figure 12. Screening colorimetric assays in the office paper platform developed: (a) paper-based sensor photograph of the colorimetric assays of hexagonal WO_3 nanoparticles at different concentrations; (b) RGB analysis of the nanoparticles, resulting in color in contact with *G. sulfurreducens* cells; (c) photograph of a single-use prototype EAB detection device, with a positive result (*G. sulfurreducens* cells in yellow and WO_3 nanoparticles in blue. The images were false colored for better understanding.); (d) RGB analysis of the color of hexagonal WO_3 nanoparticles at 15 g/L in contact with *G. sulfurreducens* cells, with the negative control *E. coli* and a blank test (results recorded after 2 hours); (e) paper-based sensor photograph used to test three different bacteria (*G. sulfurreducens* KN400, *G. sulfurreducens* DL-1 and *G. bemidjiensis*) at different growth stages (latent (1), mid-exponential (2) and exponential (3) phases); (f) RGB analysis of the colorimetric results with bacteria at exponential phase.

al. 2012), *G. sulfurreducens* DL-1, already tested and *G. bemidjiensis*, which is unable to produce current in MFCs (Nevin et al. 2005, Aklujkar et al. 2010). The results show a good discrimination between the color intensities, which indicates that the method can also be used to distinguish non-electricity and electricity production by different bacterial strains.

Table 2. Tukey's multiple comparison test for EAB detection platform.

	Mean Diff.	**q**	**Significance P<0.0001**	**Summary**	**99.9% CI of Diff.**
G. sulfurreducens versus E. coli	0.3408	211.9	Yes	***	0.3167 to 0.3649
***G. sulfurreducens versus* Blank**	0.3439	213.6	Yes	***	0.3200 to 0.3715

Therefore, the method can be further extended to screen and identify crucial proteins that participate in extracellular electron transfer processes, thus allowing further optimization in the performance of the mentioned applications.

6. Conclusion

Electrochemically active bacteria are ubiquitous in nature and have the ability to transfer electrons outside their cells—a feature that can be applicable in electricity production, which is of outmost importance in an energy-dependent world.

However, the number of isolated and identified species is still very limited and their electron transfer mechanisms are feebly understood, caressing of feasible techniques to allow the detection of these bacteria as well as to facilitate the study of their physiology and electron transfer mechanisms.

Nowadays, the most efficient screening methods are based on MFC engineering principles, though they are relatively slow (~5 to 6 days) and expensive. So, the development of rapid and simple screening methods using low-cost and available materials are today a key issue to aid in a better understanding of these types of bacteria, thus allowing further refining in the performance of these bacteria to be used in biotechnological applications.

Keywords: Electrochemically Active Bacteria, Screening Methods, Lab-on-Paper, Electrochromic Materials, Office Paper

References

Aklujkar, M., Young, N.D., Holmes, D., Chavan, M., Risso, C., Kiss, H.E., Han, C.S., Land, M.L. and Lovley, D.R. 2010. The genome of *Geobacter bemidjiensis*, exemplar for the subsurface clade of *Geobacter* species that predominate in Fe(III)-reducing subsurface environments. BMC Genomics 11: 490.

Arnold, R.G., DiChristina, T.J. and Hoffmann, M.R. 1988. Reductive dissolution of Fe(III) oxides by *Pseudomonas* sp. 200. Biotechnology and Bioengineering 32(9): 1081–1096.

Babauta, J., Renslow, R., Lewandowski, Z. and Beyenal, H. 2012. Electrochemically active biofilms: facts and fiction. A review. Biofouling 28(8): 789–812.

Bange, K. 1999. Colouration of tungsten oxide films: A model for optically active coatings. Solar Energy Materials & Solar Cells 58: 1–131.

Baron, D.B., Labelle, E., Coursolle, D., Gralnick, J.A. and Bond, D.R. 2009. Electrochemical measurement of electron transfer kinetics by *Shewanella oneidensis* MR-1. The Journal of Biological Chemistry 284(8): 28865–28873.

Biffinger, J., Ribbens, M., Ringeisen, B., Pietron, J., Finkel, S. and Nealson, K. 2009. Characterization of electrochemically active bacteria utilizing a high-throughput voltage-based screening assay. Biotechnology and Bioengineering 102(2): 436–444.

Bond, D. and Lovley, D. 2003. Electricity production by *Geobacter sulfurreducens* attached to electrodes. Applied and Environmental Microbiology 69(3): 1548–1555.

Bond, D.R., Holmes, D.E., Tender, L.M. and Lovley, D.R. 2002. Electrode-reducing microorganisms that harvest energy from marine sediments. Science 295(18): 483–485.

Bond, D.R. and Lovley, D.R. 2005. Evidence for involvement of an electron shuttle in electricity generation by Geothrix fermentans. Applied and Environmental Microbiology 71(4): 2186–2189.

Borole, A.P., Reguera, G., Ringeisen, B., Wang, Z.-W., Feng, Y. and Kim, B.H. 2011. Electroactive biofilms: Current status and future research needs. Energy & Environmental Science 4(12): 4813.

Bretschger, O., Obraztsova, A., Sturm, C., Chang, I.S., Gorby, Y., Reed, S.B., Culley, D.E., Reardon, C.L., Barua, S., Romine, M.F., Zhou, J., Beliaev, A.S., Bouhenni, R., Saffarini, D., Mansfeld, F., Kim, B.-H., Fredrickson, J.K. and Nealson, K.H. 2007. Current production and metal oxide reduction by *Shewanella oneidensis* MR-1 wild type and mutants. Applied and Environmental Microbiology 73(21): 7003–7012.

Burnes, B., Mulberry, M. and DiChristina, T. 1998. Design and application of two rapid screening techniques for isolation of Mn(IV) reduction-deficient mutants of *Shewanella putrefaciens*. Applied and Environmental Microbiology 64(7): 2716–2720.

Butler, J.E., Young, N.D., Aklujkar, M. and Lovley, D.R. 2012. Comparative genomic analysis of *Geobacter sulfurreducens* KN400, a strain with enhanced capacity for extracellular electron transfer and electricity production. BMC Genomics 13(1): 471.

Caccavo, F., Lonergan, D.J., Lovley, D.R., Davis, M., Stolz, J.F. and McInerney, M.J. 1994. *Geobacter sulfurreducens* sp. nov., a hydrogen- and acetate-oxidizing dissimilatory metal-reducing microorganism. Applied and Environmental Microbiology 60(10): 3752–3759.

Chang, I.S., Moon, H., Jang, J.K. and Kim, B.H. 2005. Improvement of a microbial fuel cell performance as a BOD sensor using respiratory inhibitors. Biosensors and Bioelectronics 20(9): 1856–1859.

Choi, G., Hassett, D.J. and Choi, S. 2015. A paper-based microbial fuel cell array for rapid and high-throughput screening of electricity-producing bacteria. The Analyst 140(12): 4277–4283.

Coates, J.D., Phillips, E.J., Lonergan, D.J., Jenter, H. and Lovley, D.R. 1996. Isolation of *Geobacter* species from diverse sedimentary environments. Applied and Environmental Microbiology 62(5): 1531–1536.

Cologgi, D.L., Lampa-Pastirk, S., Speers, A.M., Kelly, S.D. and Reguera, G. 2011. Extracellular reduction of uranium via *Geobacter* conductive pili as a protective cellular mechanism. Proceedings of the National Academy of Sciences 108(37): 15248–15252.

Costa, M.N., Veigas, B., Jacob, J.M., Santos, D.S., Gomes, J., Baptista, P.V., Martins, R., Inácio, J. and Fortunato, E. 2014. A low cost, safe, disposable, rapid and self-sustainable paper-based platform for diagnostic testing: lab-on-paper. Nanotechnology 25(9): 94006.

Deb, S.K. 2008. Opportunities and challenges in science and technology of WO_3 for electrochromic and related applications. Solar Energy Materials and Solar Cells 92(2): 245–258.

DiChristina, T.J. and DeLong, E.F. 1993. Design and application of rRNA-targeted oligonucleotide probes for the dissimilatory iron- and manganese-reducing bacterium *Shewanella putrefaciens*. Applied and Environmental Microbiology 59(12): 4152–4160.

Dinh, H.T., Kuever, J., Mubmann, M., Hassel, A.W., Stratmann, M. and Widdel, F. 2004. Iron corrosion by novel anaerobic microorganisms. Nature 427(6977): 829–832.

El-Naggar, M.Y., Wanger, G., Leung, K.M., Yuzvinsky, T.D., Southam, G., Yang, J., Lau, W.M., Nealson, K.H. and Gorby, Y. 2010. Electrical transport along bacterial nanowires from *Shewanella oneidensis* MR-1. Proceedings of the National Academy of Sciences of the United States of America 107(42): 18127–18131.

Erable, B., Duţeanu, N. and Ghangrekar, M. 2010. Application of electro-active biofilms. Biofouling 26(1): 57–71.

Fedorovich, V., Knighton, M.C., Pagaling, E., Ward, F.B., Free, A. and Goryanin, I. 2009. Novel electrochemically active bacterium phylogenetically related to *Arcobacter butzleri*, isolated from a microbial fuel cell. Applied and Environmental Microbiology 75(23): 7326–7334.

Fennessey, C.M. 2010. A Novel Mode of Bacterial Respiration: Iron Solubilization Prior to Electron Transfer. PhD Thesis, Georgia Institute of Technology, Georgia, USA.

Flemming, H.C. and Wingender, J. 2010. The biofilm matrix. Nature Reviews Microbiology 8(9): 623–633.

Fonseca, B.M., Paquete, C.M., Neto, S.E., Pacheco, I., Soares, C.M. and Louro, R.O. 2013. Mind the gap: cytochrome interactions reveal electron pathways across the periplasm of *Shewanella oneidensis* MR-1. The Biochemical Journal 449(1): 101–8.

Gorby, Y., Yanina, S., McLean, J.S., Rosso, K.M., Moyles, D., Dohnalkova, A., Beveridge, T.J., Chang, I.S., Kim, B.H., Kim, K.S., Culley, D.E., Reed, S.B., Romine, M.F., Saffarini, D.A., Hill, E.A., Shi, L., Elias, D.A., Kennedy, D.W., Pinchuk, G., Watanabe, K., Ishii, S., Logan, B., Nealson, K.H. and Fredrickson, J.K. 2006. Electrically conductive bacterial nanowires produced by *Shewanella oneidensis* strain MR-1 and other microorganisms. Proceedings of the National Academy of Sciences of the United States of America 103(30): 11358–11363.

Ha, J.-H., Muralidharan, P. and Kim, D.K. 2009. Hydrothermal synthesis and characterization of self-assembled *h*-WO_3 nanowires/nanorods using EDTA salts. Journal of Alloys and Compounds 475(1-2): 446–451.

He, H., Yuan, S.J., Tong, Z.H., Huang, Y.X., Lin, Z.Q. and Yu, H.Q. 2014. Characterization of a new electrochemically active bacterium, *Lysinibacillus sphaericus* D-8, isolated with a WO_3 nanocluster probe. Process Biochemistry 49(2): 290–294.

Hernandez, M.E. and Newman, D.K. 2001. Extracellular electron transfer. Cellular and Molecular Life Sciences: CMLS 58(11): 1562–1571.

Hernandez, M.E., Kappler, A. and Newman, D.K. 2004. Phenazines and other redox-active antibiotics promote microbial mineral reduction. Applied and Environmental Microbiology 70(2): 921–928.

Holmes, D.E., Bond, D.R., O'Neil, R.A., Reimers, C.E., Tender, L.R. and Lovley, D.R. 2004. Microbial communities associated with electrodes harvesting electricity from a variety of aquatic sediments. Microbial Ecology 48(2): 178–190.

Holmes, D.E., Nicoll, J.S., Bond, D.R. and Lovley, D.R. 2004. Potential role of a novel psychrotolerant member of the family *Geobacteraceae, Geopsychrobacter electrodiphilus* gen. nov., sp. nov., in electricity production by a marine sediment fuel cell. Applied and Environmental Microbiology 70: 6023–6030.

Hou, H., Li, L., Cho, Y., Figueiredo, P. and Han, A. 2009. Microfabricated microbial fuel cell arrays reveal electrochemically active microbes. PloS One 4(8): 6570.

Hou, H., Li, L., Ceylan, C.Ü., Haynes, A., Cope, J., Wilkinson, H.H., Erbay, C., Figueiredo, P. de and Han, A. 2012. A microfluidic microbial fuel cell array that supports long-term multiplexed analyses of electricigens. Lab on a Chip 12(20): 4151.

Inoue, K., Qian, X., Morgado, L., Kim, B.-C., Mester, T., Izallalen, M., Salgueiro, C.A. and Lovley, D.R. 2010. Purification and characterization of OmcZ, an outer-surface, octaheme *c*-type cytochrome essential for optimal current production by *Geobacter sulfurreducens*. Applied and Environmental Microbiology 76(12): 3999–4007.

Jiang, X., Hu, J., Fitzgerald, L., Biffinger, J.C., Xie, P., Ringeisen, B.R. and Lieber, C.M. 2010. Probing electron transfer mechanisms in *Shewanella oneidensis* MR-1 using a nanoelectrode platform and single-cell imaging. Proceedings of the National Academy of Sciences of the United States of America 107(39): 16806–16810.

Jokerst, J.C., Adkins, J., Bisha, B., Mentele, M.M., Goodridge, L.D. and Henry, C.S. 2012. Development of a paper-based analytical device for colorimetric detection of select foodborne pathogens. Analytical Chemistry 84(6): 2900–7.

Khan, M.M., Kalathil, S., Lee, J. and Cho, M.H. 2012. Synthesis of cysteine capped silver nanoparticles by electrochemically active biofilm and their antibacterial activities. Bulletin of the Korean Chemical Society 33(8): 2592–2596.

Kharade, R.R., Patil, K.R., Patil, P.S. and Bhosale, P.N. 2012. Novel microwave assisted sol–gel synthesis (MW-SGS) and electrochromic performance of petal like *h*-WO_3 thin films. Materials Research Bulletin 47(7): 1787–1793.

Kim, B.H., Ikeda, T., Park, H.S., Kim, H.J., Hyun, M.S., Kano, K., Takagi, K. and Tatsumi, H. 1999. Electrochemical activity of an Fe(III)-reducing bacterium, *Shewanella putrefaciens* IR-1, in the presence of alternative electron acceptors. Biotechnology Techniques 13(7): 475–478.

Kim, B.H., Chang, I.S., Gil, G.C., Park, H.S. and Kim, H.J. 2003. Novel BOD (biological oxygen demand) sensor using mediator-less microbial fuel cell. Biotechnology Letters 25(7): 541–545.

Kim, G.T., Hyun, M.S., Chang, I.S., Kim, H.J., Park, H.S., Kim, B.H., Kim, S.D., Wimpenny, J.W.T. and Weightman, A.J. 2005. Dissimilatory Fe(III) reduction by an electrochemically active lactic acid bacterium phylogenetically related to *Enterococcus gallinarum* isolated from submerged soil. Journal of Applied Microbiology 99(4): 978–987.

Kim, G.T., Webster, G., Wimpenny, J.W.T., Kim, B.H., Kim, H.J. and Weightman, A.J. 2006. Bacterial community structure, compartmentalization and activity in a microbial fuel cell. Journal of Applied Microbiology 101(3): 698–710.

Kim, H.J., Park, H.S., Hyun, M.S., Chang, I.S., Kim, M. and Kim, B.H. 2002. A mediator-less microbial fuel cell using a metal reducing bacterium, *Shewanella putrefaciens*. Enzyme and Microbial Technology 30(2): 145–152.

Kim, M., Sik Hyun, M., Gadd, G.M. and Joo Kim, H. 2007. A novel biomonitoring system using microbial fuel cells. Journal of Environmental Monitoring: JEM 9(12): 1323–1328.

Kouzuma, A., Kasai, T., Nakagawa, G., Yamamuro, A., Abe, T. and Watanabe, K. 2013. Comparative metagenomics of anode-associated microbiomes developed in rice paddy-field microbial fuel cells. PloS One 8(11): e77443.

Larsen, L.H., Damgaard, L.R., Kjaer, T., Stenstrøm, T., Lynggaard-Jensen, A. and Revsbech, N.P. 2000. Fast responding biosensor for on-line determination of nitrate/nitrite in activated sludge. Water Research 34(9): 2463–2468.

Lee, S.-H., Deshpande, R., Parilla, P.A., Jones, K.M., To, B., Mahan, A.H. and Dillon, A.C. 2006. Crystalline WO_3 nanoparticles for highly improved electrochromic applications. Advanced Materials 18(6): 763–766.

Levar, C.E., Chan, C.H., Mehta-Kolte, M.G. and Bond, D.R. 2014. An inner membrane cytochrome required only for reduction of high redox potential extracellular electron acceptors. mBio 5(6): e02034.

Liana, D.D., Raguse, B., Gooding, J.J. and Chow, E. 2012. Recent advances in paper-based sensors. Sensors 12(12): 11505–11526.

Lies, D.P., Hernandez, M.E., Kappler, A., Mielke, R.E., Gralnick, J. and Newman, D.K. 2005. *Shewanella oneidensis* MR-1 uses overlapping pathways for iron reduction at a distance and by direct contact under conditions relevant for biofilms. Applied and Environmental Microbiology 71(8): 4414–26.

Liu, W., Huang, S., Zhou, A., Zhou, G., Ren, N., Wang, A. and Zhuang, G. 2012. Hydrogen generation in microbial electrolysis cell feeding with fermentation liquid of waste activated sludge. International Journal of Hydrogen Energy 37(18): 13859–13864.

Logan, B.E. 2009. Exoelectrogenic bacteria that power microbial fuel cells. Nature Reviews. Microbiology 7(5): 375–81.

Lovley, D.R. and Phillips, E.J. 1988. Novel mode of microbial energy metabolism: organic carbon oxidation coupled to dissimilatory reduction of iron or manganese. Applied and Environmental Microbiology 54(6): 1472–1480.

Lovley, D.R., Holmes, D.E. and Nevin, K.P. 2004. Dissimilatory Fe(III) and Mn(IV) reduction. Advances in Microbial Physiology 49(2): 219–86.

Lovley, D.R. 2006a. Bug juice: harvesting electricity with microorganisms. Nature Reviews. Microbiology 4(7): 497–508.

Lovley, D.R. 2006b. Microbial energizers: Fuel cells that keep on going. Microbe. 1(7): 323–329.

Lovley, D.R. 2008. The microbe electric: conversion of organic matter to electricity. Current Opinion in Biotechnology 19(6): 564–571.

Lovley, D.R., Ueki, T., Zhang, T., Malvankar, N.S., Shrestha, P.M., Flanagan, K., Aklujkar, M., Butler, J.E., Giloteaux, L., Rotaru, A.-E., Holmes, D.E., Franks, A.E., Orellana, R., Risso, C. and Nevin, K.P. 2011. *Geobacter:* the microbe electric's physiology, ecology, and practical applications. Advances in Microbial Physiology 59: 1–100.

Luijten, M.L.G.C., Roelofsen, W., Langenhoff, A.A.M., Schraa, G. and Stams, A.J.M. 2004. Hydrogen threshold concentrations in pure cultures of halorespiring bacteria and at sites polluted with chlorinated ethenes. Environmental Microbiology 6(6): 646–650.

Marques, A.C., Santos, L., Costa, M.N., Dantas, J.M., Duarte, P., Gonçalves, A., Martins, R., Salgueiro, C.A. and Fortunato, F. 2015. Office Paper Platform for Bioelectrochromic Detection of Electrochemically Active Bacteria using Tungsten Oxide Nanoprobes. Scientific Reports 5(9910): 1–7.

Marsili, E., Rollefson, J.B., Baron, D.B., Hozalski, R.M. and Bond, D.R. 2008. Microbial biofilm voltammetry: Direct electrochemical characterization of catalytic electrode-attached biofilms. Applied and Environmental Microbiology 74(23): 7329–7337.

Martinez, A., Phillips, S., Whitesides, G. and Carrilho, E. 2010. Diagnostics for the Developing World: Microfluidic Paper-Based Analytical Devices. Analytical Chemistry 82: 3–10.

Martinez, A.W., Phillips, S.T., Butte, M.J. and Whitesides, G.M. 2007. Patterned paper as a platform for inexpensive, low-volume, portable bioassays. Angewandte Chemie (International ed. in English) 46(8): 1318–1320.

Mathuriya, S.A. and Sharma, V.N. 2010. Bioelectricity production from various wastewaters through microbial fuel cell technology. Journal of Biochemical Technology 2(2009): 133–137.

Morgado, L., Bruix, M., Pessanha, M., Londer, Y.Y. and Salgueiro, C.A. 2010. Thermodynamic characterization of a triheme cytochrome family from *Geobacter sulfurreducens* reveals mechanistic and functional diversity. Biophysical Journal 99(1): 293–301.

Myers, C.R. and Nealson, K.H. 1988. Bacterial manganese reduction and growth with manganese oxide as the sole electron acceptor. Science 240(4857): 1319–1321.

Nevin, K.P. and Lovley, D.R. 2002. Mechanisms for Fe(III) Oxide reduction in sedimentary environments. Geomicrobiology Journal 19(2): 141–159.

Nevin, K.P., Holmes, D.E., Woodard, T.L., Hinlein, E.S., Ostendorf, D.W. and Lovley, D.R. 2005. *Geobacter bemidjiensis* sp. nov. and *Geobacter psychrophilus* sp. nov., two novel Fe(III)-reducing subsurface isolates. International Journal of Systematic and Evolutionary Microbiology 55(4): 1667–1674.

Paquete, C.M., Fonseca, B.M., Cruz, D.R., Pereira, T.M., Pacheco, I., Soares, C.M. and Louro, R.O. 2014. Exploring the molecular mechanisms of electron shuttling across the microbe/metal space. Frontiers in Microbiology 5(JUN): 1–12.

Park, H.S., Kim, B.H., Kim, H.J.H.S., Kim, G.T., Kim, M., Chang, I.S., Park, Y.K., and Chang, H.I. 2001. A novel electrochemically active and Fe(III)-reducing bacterium phylogenetically related to *Clostridium butyricum* isolated from a microbial fuel cell. Anaerobe 7(6): 297–306.

Pelton, R. 2009. Bioactive paper provides a low-cost platform for diagnostics. TrAC Trends in Analytical Chemistry 28(8): 925–942.

Pham, C.A., Jung, S.J., Phung, N.T., Lee, J., Chang, I.S., Kim, B.H., Yi, H. and Chun, J. 2003. A novel electrochemically active and Fe(III)-reducing bacterium phylogenetically related to *Aeromonas hydrophila*, isolated from a microbial fuel cell. FEMS Microbiology Letters 223(1): 129–134.

Pokkuluri, P.R., Londer, Y.Y., Duke, N.E.C., Pessanha, M., Yang, X., Orshonsky, V., Orshonsky, L., Erickson, J., Zagyanskiy, Y., Salgueiro, C.A. and Schiffer, M. 2011. Structure of a novel dodecaheme cytochrome *c* from *Geobacter sulfurreducens* reveals an extended 12 nm protein with interacting hemes. Journal of Structural Biology 174(1): 223–33.

Qian, X., Mester, T., Morgado, L., Arakawa, T., Sharma, M.L., Inoue, K., Joseph, C., Salgueiro, C.A., Maroney, M.J. and Lovley, D.R. 2011. Biochemical characterization of purified OmcS, a *c*-type cytochrome required for insoluble Fe(III) reduction in *Geobacter sulfurreducens*. Biochimica et Biophysica Acta 1807(4): 404–12.

Rabaey, K., Boon, N., Siciliano, S.D., Verstraete, W. and Verhaege, M. 2004. Biofuel cells select for microbial consortia that self-mediate electron transfer. Applied and Environmental Microbiology 70(9): 5373–5382.

Rabaey, K., Boon, N., Höfte, M. and Verstraete, W. 2005. Microbial phenazine production enhances electron transfer in biofuel cells. Environmental Science and Technology 39(9): 3401–3408.

Rabaey, K. and Rozendal, R.A. 2010. Microbial electrosynthesis—revisiting the electrical route for microbial production. Nature Reviews Microbiology 8(10): 706–716.

Ray, R., Lizewski, S., Fitzgerald, L.A., Little, B. and Ringeisen, B.R. 2010. Methods for imaging *Shewanella oneidensis* MR-1 nanofilaments. Journal of Microbiological Methods 82(2): 187–91.

Regan, J. and Logan, B. 2006. Microbial challenges and fuel cell aplications. Environmental Science & Technology 1 (September): 5172–80.

Reguera, G., McCarthy, K.D., Mehta, T., Nicoll, J.S., Tuominen, M.T. and Lovley, D.R. 2005. Extracellular electron transfer via microbial nanowires. Nature 435(7045): 1098–10101.

Reguera, G., Nevin, K.P., Nicoll, J.S., Covalla, S.F., Woodard, T.L. and Lovley, D.R. 2006. Biofilm and nanowire production leads to increased current in *Geobacter sulfurreducens* fuel cells. Applied and Environmental Microbiology 72(11): 7345–7348.

Richter, H., Lanthier, M., Nevin, K.P. and Lovley, D.R. 2007. Lack of electricity production by *Pelobacter carbinolicus* indicates that the capacity for Fe(III) oxide reduction does not necessarily confer electron transfer ability to fuel cell anodes. Applied and Environmental Microbiology 73(16): 5347–5353.

Ringeisen, B.R., Lizewski, S.E., Fitzgerald, L., Biffinger, J.C., Knight, C.L., Crookes-Goodson, W.J. and Wu, P.K. 2010. Single cell isolation of bacteria from microbial fuel cells and potomac river sediment. Electroanalysis 22(7-8): 875–882.

Santos, L., Neto, J.P., Crespo, A., Nunes, D., Costa, N., Fonseca, I.M., Barquinha, P., Pereira, L., Silva, J., Martins, R. and Fortunato, E. 2014. WO_3 nanoparticle-based conformable pH sensor. ACS Applied Materials & Interfaces 13(6): 12226–12234.

Santos, L., Silveira, C.M., Elangovan, E., Neto, J.P., Nunes, D., Pereira, L., Martins, R., Viegas, J., Moura, J.J.G., Todorovic, S., Almeida, M.G. and Fortunato, E. 2016. Synthesis of WO_3 nanoparticles for biosensing applications. Sensors and Actuators B: Chemical 223: 186–194.

Santos, T.C., Silva, M.A., Morgado, L., Dantas, J.M. and Salgueiro, C.A. 2015. Diving into the redox properties of *Geobacter sulfurreducens* cytochromes: a model for extracellular electron transfer. Dalton transactions (Cambridge, England: 2003) 44(20): 9335–44.

Seidel, J., Hoffmann, M., Ellis, K.E., Seidel, A., Spatzal, T., Gerhardt, S., Elliott, S.J. and Einsle, O. 2012. MacA is a second cytochrome *c* peroxidase of *Geobacter sulfurreducens*. Biochemistry 51(13): 2747–56.

Srikanth, S., Marsili, E., Flickinger, M.C. and Bond, D.R. 2008. Electrochemical characterization of *Geobacter sulfurreducens* cells immobilized on graphite paper electrodes. Biotechnology and Bioengineering 99(5): 1065–1073.

Srikanth, S., Venkateswar Reddy, M. and Venkata Mohan, S. 2012. Microaerophilic microenvironment at biocathode enhances electrogenesis with simultaneous synthesis of polyhydroxyalkanoates (PHA) in bioelectrochemical system (BES). Bioresource Technology 125: 291–299.

Stams, A.J.M., de Bok, F.M., Plugge, C.M., van Eekert, M.H., Dolfing, J. and Schraa, G. 2006. Exocellular electron transfer in anaerobic microbial communities. Environmental Microbiology 8(3): 371–82.

Stolz, J.F. and Oremland, R.S. 1999. Bacterial respiration of arsenic and selenium. FEMS Microbiology Rev. 23(5): 615–627.

Strycharz-Glaven, S.M., Snider, R.M., Guiseppi-Elie, A. and Tender, L.M. 2011. On the electrical conductivity of microbial nanowires and biofilms. Energy & Environmental Science 4(11): 4366–4379.

Tartakovsky, B., Manuel, M.F., Wang, H. and Guiot, S.R. 2009. High rate membrane-less microbial electrolysis cell for continuous hydrogen production. International Journal of Hydrogen Energy 34(2): 672–677.

Veigas, B., Jacob, J.M., Costa, M.N., Santos, D.S., Viveiros, M., Inácio, J., Martins, R., Barquinha, P., Fortunato, E. and Baptista, P.V. 2012. Gold on paper-paper platform for Au-nanoprobe TB detection. Lab on a Chip 12(22): 4802–8.

Venkata Mohan, S., Velvizhi, G., Vamshi Krishna, K. and Lenin Babu, M. 2014. Microbial catalyzed electrochemical systems: A bio-factory with multi-facet applications. Bioresource Technology 165: 355–364.

Von Canstein, H., Ogawa, J., Shimizu, S. and Lloyd, J.R. 2008. Secretion of flavins by *Shewanella* species and their role in extracellular electron transfer. Applied and Environmental Microbiology 74(3): 615–623.

Yi, H., Nevin, K.P., Kim, B.-C., Franks, A.E., Klimes, A., Tender, L.M. and Lovley, D.R. 2009. Selection of a variant of *Geobacter sulfurreducens* with enhanced capacity for current production in microbial fuel cells. Biosensors and Bioelectronics 24(12): 3498–3503.

Yuan, S.-J., He, H., Sheng, G.-P., Chen, J.-J., Tong, Z.-H., Cheng, Y.-Y., Li, W.-W., Lin, Z.-Q., Zhang, F. and Yu, H.-Q. 2013. A photometric high-throughput method for identification of electrochemically active bacteria using a WO_3 nanocluster probe. Scientific Reports 3: 1315.

Zacharoff, L., Chan, C.H. and Bond, D.R. 2016. Reduction of low potential electron acceptors requires the CbcL inner membrane cytochrome of *Geobacter sulfurreducens*. Bioelectrochemistry 107: 7–13.

Zhang, Y. and Rochefort, D. 2013. Fast and effective paper based sensor for self-diagnosis of bacterial vaginosis. Analytica Chimica. Acta 800: 87–94.

Zheng, H., Ou, J.Z., Strano, M.S., Kaner, R.B., Mitchell, A. and Kalantar-zadeh, K. 2011. Nanostructured tungsten oxide—Properties, synthesis, and applications. Advanced Functional Materials 21(12): 2175–2196.

Zhou, M., Yang, M. and Zhou, F. 2014. Paper based colorimetric biosensing platform utilizing cross-linked siloxane as probe. Biosensors & Bioelectronics 55: 39–43.

Zuo, Y., Xing, D., Regan, J.M. and Logan, B.E. 2008. Isolation of the exoelectrogenic bacterium *Ochrobactrum anthropi* YZ-1 by using a U-tube microbial fuel cell. Applied and Environmental Microbiology 74(10): 3130–7.

4

Application of Quantitative Real-time PCR for Microbial Community Analysis in Environmental Research

Sevcan Aydin

1. Introduction

Anton Von Leeuwenhoek and Robert Hook discovered the first microorganism in the seventeenth century when their common interests in microscopes brought them to work together on a groundbreaking project. Since then, researchers have continued to invest significant resources in studying the ways in which microorganisms interact with their environment (Amann et al. 1995, 2001, Su et al. 2012). Culture-based methods are commonly employed as a means of investigating the microbial ecology of both natural, untouched environments and those that have been anthropogenically altered by human activity. However, gaining in-depth insights into microorganisms can be extremely challenging because many groups of interest to scientists cannot be cultured in a laboratory setting and have been detected through the use of culture-independent approaches (Nagarajan and Loh 2014, Aydin 2016). Previous attempts to

Department of Genetics and Bioengineering, Nişantaşı University, Maslak, 34469, Istanbul, Turkey.
E-mail: sevcan_aydn@hotmail.com

cultivate environmental communities have only successfully developed less than one per cent of the total prokaryotic species present in the sample (Madsen 2005). Furthermore, scientists have demonstrated bias in their assessment of microbial genetic diversity, with certain populations of microorganisms being clearly favored over others (Vanwonterghem et al. 2014, Baldrian et al. 2014, Aydin 2016). Molecular methods that involve the isolation and assessment of DNA, RNA, proteins, metabolites and stable/radioactive isotopes from environmental samples have been successfully deployed and these can provide valuable insights into the structure and functional behavior of microbial communities, as seen in Fig. 1. Both, entire genomes and selected genes, can be analyzed using a culture-independent nucleic acid approach that is based on comparative analyses of rRNA. According to an analysis of these rRNA signatures, scientists have segmented cellular life according to three primary classifications: Eukarya, Bacteria and Archaea (Abbasi-Guendouz et al. 2013, Nagarajan et al. 2014).

In more recent times, techniques such as polymerase chain reaction (PCR) and Sanger sequencing have allowed scientists to better comprehend the phylogenetic and functional characteristics of microorganisms. Techniques that operate by extracting nucleic acids rely on enzymatic amplification of certain genes from the complex genomic DNA of environmental samples

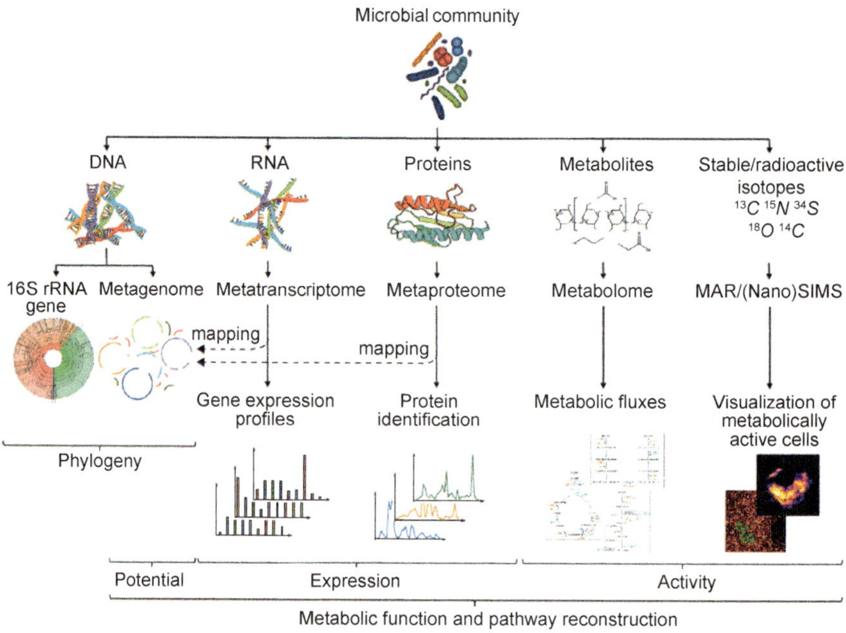

Figure 1. A combination of molecular methods from environmental samples (Vanwonterghem et al. 2014).

(Su et al. 2012, Nagarajan and Loh 2014, Aydin 2016). The understanding of molecular approaches and comprehension of the way in which they interact with the environment has also been aided by techniques, such as quantitative real-time PCR (Zarraonaindia et al. 2013, Tang et al. 2014). It is also possible to quantify different nucleic acid sequences by employing a real-time quantitative polymerase chain reaction (qPCR) analysis method, which works by investigating the concentration of specific nucleic acid sequences as they are enzymatically amplified *in vitro*. Scientists have discovered that the higher the initial concentration of the target gene, the less time is takes for it to reach a set threshold concentration (Yu et al. 2006, Smith and Osborn 2009).

Environmental microbiology as a field of study has seen significant developments in recent years as a result of the introduction of new molecular genomic tools. It is now much easier to observe the structural and functional diversity of environmental microbial communities using quantitative real-time PCR approaches. The main objective of this chapter is to examine quantitative real-time PCR techniques that are available in order to identify the advantages and disadvantages due to the use of the most common molecular biology techniques that are currently applied to examine environmental samples for research purposes.

2. Quantitative Real-time PCR Approaches in Environmental Research

qPCR is a considerably essential approach to detect phylogenetic and functional gene changes in various environmental or experimental conditions across temporal and spatial scales. Variations in gene abundances and gene expression levels can be compared with differences in abiotic or biotic features and/or biological activities and process rates through the use of any quantitative data that is produced through this process (Kim et al. 2013, Aydin et al. 2015a,b, Aydin 2016). The arrangement of qPCR data sets, which are described by the abundance of specific bacteria or genes for completion of other quantitative environmental data sets, has significance in microbial ecology for comprehension of the functions and influences of certain microbial and functional groups within the environment. Reverse transcription (RT)-qPCR assays are also offering a specific method for quantifying gene expression and relationships with microbial activity to function.

There are two main methods for qPCR—SYBR green (a non-specific fluorogenic molecule which binds to double-stranded DNA) and a dual-labeled TaqMan probe (binding with specific DNA target) as seen in Fig. 2.

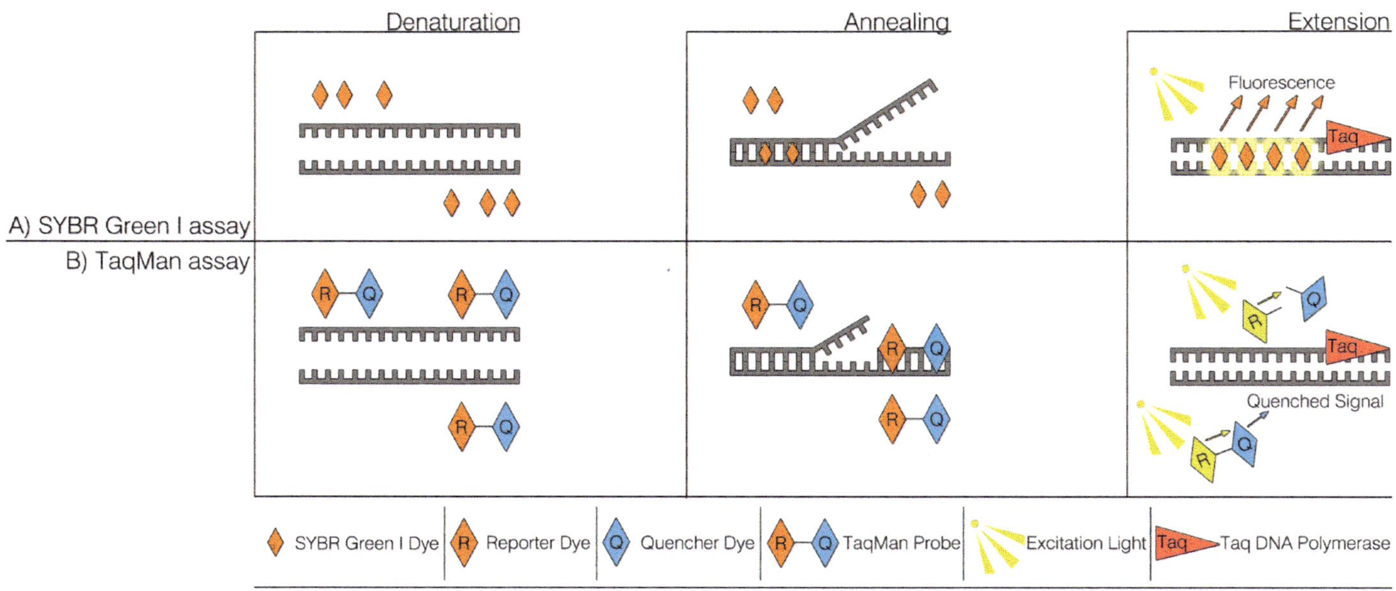

Figure 2. SYBR Green I and TaqMan assays for qPCR detection.

Among the several detection chemistries, SYBR Green I and TaqMan assays are most extensively used approaches in environmental studies. SYBR Green I is a dye, which binds to the minor groove of double-stranded DNA in a sequence-independent way, emitting 1000-fold greater fluorescence than when unbound, after each amplification cycle is quantified by the relative fluorescent intensity of each sample and the amount of DNA produced (Kim et al. 2013). In TaqMan, binding of target DNA and the dual labeled probe is expected along with the standard PCR amplification primers. The probe is degraded into individual nucleotides caused by the 5'-3' exonuclease activity of Taq polymerase during the primer extension phase (Holland et al. 1991, Wittwer et al. 1997, Giulietti et al. 2001). The fluorescence can be detected with no more quenching of the fluorescent marker. The need for a probe complicates the design process of primer/probe sets particular to the sequence of interest: three oligonucleotides specific to different regions in one target sequence of limited length. Such complications can lead to the design and use of flawed primer/probe sets, which can produce distorted results (Kim et al. 2013). For both of them, determination of cycle threshold (Ct) or a value called the critical is the relative cycle at which the fluorescence of a sample increased above background. By comparing cycle threshold values of samples with unknown amounts of initial target DNA to those of standards with known starting quantities of template DNA, it is possible to correctly quantify the plenty copies of a particular gene sequence in a mixed community DNA samples (Smith and Osborn 2009).

3. Quantification of Microbial Communities and Gene Expression in Environmental Research

Unlike traditional PCR, which is based on end-point detection of amplified genes, q-PCR uses fluorescence-based detection, such as SYBR Green or fluorescent probes (TaqMan probe), to measure the accumulation of amplicons in real time during each cycle of the PCR (Kim et al. 2013). Lee et al. (2009) examined changes in community structures of methanogens and their effects on the reactor performance in anaerobic digesters, using 16S rRNA gene-specific q-PCR assays. According to the results, *Methanosarcinaceae* and *Methanomicrobiales* were the dominant methanogen groups. Similarly, Shin et al. (2011) investigated methanogen communities in two anaerobic digesters. The q-PCR results, which are based on the 16S rRNA, show the domination of *Methanosarcinales* on all methanogenic communities in 6 samples where chemical oxygen demand (COD) was higher than 50 per cent. However, *Methanobacteriales* dominated on the other two samples with 20 per cent COD removal efficiency. Yu et al. (2005) have also designed and tested the data set of TaqMan qPCR whose primers and probes are directed to the four primary orders of methanogens witnessed in anaerobic environments. Then, the dynamics of acetoclastic methanogens are monitored by various acetate concentrations in anaerobic digestion systems. Likewise, Franke-Whittle et al. (2009) have designed a refined set of primers using SYBR green-based assays for quantifying the genera *Methanoculleus*, *Methanosarcina* and *Methanothermobacter*. Studies that are based on 16S rRNA are used to evaluate the microbial diversity of the anaerobic environment. Universal bacterial primers are used to amplify hypervariable 16S rRNA regions. The nucleotide barcodes are attached before pyrosequencing in this type of analysis. This method bears similarities to the traditional cloning-and-sequencing protocol of early metagenomic studies, but differs on larger scales. Pyrosequencing has a major advantage over the 16S rRNA cloning method as it can produce numerous sequencing runs without requiring the creation of a clone library. Beside the targeting 16S rRNA, qPCR was also applied to functional genes as an alternative method for enumerating methanogens. Cetecioglu et al. (2016) also performed two lab-scale anaerobic sequencing batch reactor (ASBR) tests in order to determine microbial community dynamics. While synthetic pharmaceutical industry wastewater with sulfamethoxazole (SMX) was used in the test reactor, sulfamethoxazole was not used in the control reactor. Microbial community changes in the ASBRs were revealed by qPCR analysis. It was found that there is an important reduction in the bacterial community by approximately 84 per cent. However, it was shown that there is an increase in methanogens by roughly 97 per cent through the operation. As a result of qPCR analysis, it was shown that there is an important correlation between microbial community and the reactor operation data.

qPCR is also often used in environmental research for determining gene and transcript numbers that are present within environmental samples (Shahi et al. 2016). qPCR assay is determined for the target specificity with the design of the primers, allowing quantification of taxonomic or functional gene markers present within a microbial community from the domain level down to the quantification of individual species or phylotypes. RT-qPCR assays, which are also connected Reverse transcription (RT) analyses, and qPCR methods offer a strong tool for quantifying gene expression and relationships with biological activity to ecological function. For example, Aydin et al. (2015a) used quantitative real-time PCR in order to detect the impact of different antibiotic combinations. In this study, the inhibition level on the gene expression of microorganisms was also evaluated by three primer sets. It was shown that homoacetogens, methanogens and acetoclastic methanogens have a main role to play in the anaerobic degradation of antibiotics. In stability of microorganisms in an anaerobic reactor can cause a decrease in the performance of anaerobic systems. In order to comprehend how triclosan, which is a common antimicrobial agent, influences the microbial community dynamics, functional and resistance genes in anaerobic digesters, McNamara et al. (2014) studied municipal wastewater treatment plants. Microbial community composition dependence on triclosant antibiotics was investigated using the qPCR method. As a consequence of the study, it was indicated that microbial community structure and exposure history change the effect of triclosan. Additionally, Xu et al. (2015) determined 13 antibiotic resistance genes (ARGs) (6 tet genes: *tetA*, *tetB*, *tetE*, *tetW*, *tetM* and *tetZ*, 3 sulfonamide genes: *sul1*, *sul2* and *sul3*, and 4 quinolone genes: *gryA*, *parC*, *qnrC* and *qnrD*) through quantitative PCR assays. The results indicate that sewage has a wide range of resistance genes and which must be treated before discharge into the natural water body. Birošová et al. (2014) also aimed at detecting the occurrence and levels of 33 antibiotics and antibiotic-resistant microorganisms in two wastewater treatment systems in Bratislava, Slovakia. They compared a influent and effluent wastewater for monitoring the differences in antibiotic concentrations that are affected by seasonal differences due to antibiotic consumption and antibiotic resistance of coliforms and streptococci in sewage sludge. Antibiotic-resistant strains were detected by bacteriological counting. Likewise, an anaerobic sequencing batch reactor was used by Aydin et al. (2015b) so as to treat pharmaceutical wastewater contaminated with combinations of sulfamethoxazole-tetracycline-erythromycin (STE) and sulfamethoxazole-tetracycline (ST). The main purpose of the study was to analyze the presence and variation in the antibiotic-resistance genes. It was also aimed to study the impact of the combination of antibiotics on the improvement and spread of ARGs in anaerobic sequencing batch reactors. For this purpose, qPCR assays were performed and effects of different concentrations of antibiotic mixtures for progress of ARGs were determined. Illumina sequencing, which is a molecular method of next generation sequencing, was also used on the sludge and effluent of the STE and ST reactors owing to the limitation of qPCR primers to determine ARGs. According to these results, next generation sequencing analyses show a great potential to overcome limitation of qPCR and possibility of bias in the amplification process. These studies also indicate that the use of Illumina sequencing provides an in-depth analysis of the ARGs in water and sludge examples. Taken together, these results suggest that by using the next generation sequencing to achieve high sequencing depth, it is possible to control the changes in the structure of a microbial community and genes with a more in-depth understanding that successfully facilitates biogas production. On the other hand, next generation analyses still remain costly as compared to the qPCR approach. Omics (metagenomics, metaproteomics, proteogenomics and metatranscriptomics) approaches represent quicker and more efficient methods of developing a detailed understanding of active and total microbial community and identifying novel enzymes than conventional genomics and transcriptomics. But, they result in a large amount of data, which can be challenging to interpret (Baldrian and López-Mondéjar 2014). As such, these approaches need to be complemented with computational methods that allow the interpretation of the data generated in order to effectively identify and encode enzymes that are of interest in the various fields of application.

4. Conclusion

While significant advancements have been made with regard to the classification of microbial populations in an environment using the qPCR approach, there is still a lack of understanding of the functional behavior of uncultured organisms. Improvement of knowledge in this area remains an intimidating a task because the majority of genes that have been identified thus far have no homologous representatives in the existing databases. The application of new microbial sequencing techniques, such as sequencing analyses, have improved understanding; however, a number of significant technical challenges remain. In order to improve the effectiveness and depth of study results, scientists need to apply an arrangement of several methods in their quest to understand the variety, function and ecology of microbial assemblages.

Keywords: Polymerase chain reaction (PCR), quantitative PCR (qPCR), real time PCR, environmental monitoring

References

Abbassi-Guendouz, A., Trably, E., Hamelin, J., Dumas, C., Steyer, J.P., Delgenès, J.P. and Escudié, R. 2013. Microbial community signature of high-solid content methanogenic ecosystems. Bioresource Technology 133: 256–262.

Amann, R.I., Ludwig, W. and Schleifer, K.H. 1995. Phylogenetic identification and *in situ* detection of individual microbial cells without cultivation. Microbiological Reviews 59: 143–169.

Amann, R.I., Fuchs, B.M. and Behrens, S. 2001. The identification of microorganisms by fluorescence *in situ* Hybridization. Current Opinion in Biotechnology 12: 231–236.

Aydin, S., Ince, B. and Ince, O. 2015a. Application of real-time PCR to determination of combined effect of antibiotics on bacteria, methanogenic archaea, archaea in anaerobic sequencing batch reactors. Water Research 76: 88–98.

Aydin, S., Ince, B. and Ince, O. 2015b. Development of antibiotic resistance genes in microbial communities during long-term operation of anaerobic reactors in the treatment of pharmaceutical wastewater. Water Research 83: 337–344.

Aydin, S. 2016. Microbial sequencing methods for monitoring of anaerobic treatment of antibiotics to optimize performance and prevent system failure. Applied Microbiology and Biotechnology 100(12): 5313–5321.

Baldrian, P. and López-Mondéjar, R. 2014. Microbial genomics, transcriptomics and proteomics: new discoveries in decomposition research using complementary methods. Applied Microbiology and Biotechnology 98: 1531–1537.

Birošová, L., Mackuľak, T., Bodík, I., Ryba, J., Škubák, J. and Grabic, R. 2014. Pilot study of seasonal occurrence and distribution of antibiotics and drug resistant bacteria in wastewater treatment plants in Slovakia. Science of the Total Environment 490: 440–444.

Cetecioglu, Z., Ince, B., Orhon, D. and Ince, O. 2016. Anaerobic sulfamethoxazole degradation is driven by homoacetogenesis coupled with hydrogenotrophic methanogenesis. Water Research 90: 79–89.

Franke-Whittle, I.H., Goberna, M. and Insam, H. 2009. Design and testing of real-time PCR primers for the quantification of Methanoculleus, Methanosarcina, Methanothermobacter and a group of uncultured methanogens. Can. J. Microbiol. 55: 611–616

Giulietti, A., Overbergh, L., Valckx, D., Decallonne, B., Bouillon, R. and Mathieu, C. 2001. An overview of real-time quantitative PCR: Applications to quantify cytokine gene expression. Methods 25: 386–401.

Holland, P.M., Abramson, R.D., Watson, R. and Gelfand, D.H. 1991. Detection of specific polymerase chain reaction product by utilizing the 50–30 exonuclease activity of Thermus aquaticus DNA polymerase. P. Natl. Acad. Sci. USA 88: 7276–7280.

Kim, J., Lim, J. and Lee, C. 2013. Quantitative real-time PCR approaches for microbial community studies in wastewater treatment systems: Applications and considerations. Biotechnology Advances 31: 1358–1373.

Lee, C., Kim, J., Hwang, K., O'Flaherty, V. and Hwang, S. 2009. Quantitative analysis of methanogenic community dynamics in three anaerobic batch digesters treating different wastewaters. Water Research 43: 157–65.

Madsen, E.L. 2005. Identifying microorganisms responsible for ecologically significant biogeochemical processes. Nature Rev. 3: 439–446.

McNamara, P.J., LaPara, T.M. and Novak, P.J. 2014. The impacts of triclosan on anaerobic community structures, function, and antimicrobial resistance. Environmental Science & Technology 48: 7393–7400.

Nagarajan, K. and Loh, K.C. 2014. Molecular biology-based methods for quantification of bacteria in mixed culture: perspectives and limitations. Applied Microbiology and Biotechnology 98: 6907–6919.

Shahi, A., Aydin, S., Ince, B. and Ince, O. 2016. Evaluation of microbial population and functional genes during the bioremediation of petroleum-contaminated soil as an effective monitoring approach. Ecotoxicology and Environmental Safety 125: 153–160.

Shin, S.G., Zhou, B.W., Lee, S., Kim, W. and Hwang, S. 2011. Variations in methanogenic population structure under overloading of pre-acidified high-strength organic wastewaters. Process Biochemistry 46: 1035–8.

Smith, C.J. and Osborn, A.M. 2009. Advantages and limitations of quantitative PCR (qPCR)-based approaches in microbial ecology. FEMS Microbiology Ecology 67(1): 6–20.

Su, C., Lei, L., Duan, Y., Zhang, K.Q. and Yang, J. 2012. Culture-independent methods for studying environmental microorganisms: Methods, application, and perspective. Applied Microbiology and Biotechnology 93: 993–1003.
Tang, X., Mu, X., Shao, H., Wang, H. and Brestic, M. 2014. Global plant-responding mechanisms to salt stress: physiological and molecular levels and implications in biotechnology. Critical Reviews in Biotechnology 5: 1–13.
Vanwonterghem, I., Jensen, P.D., Ho, D.P., Batstone, D.J. and Tyson, G.W. 2014. Linking microbial community structure, interactions and function in anaerobic digesters using new molecular techniques. Current Opinion in Biotechnology 27: 55–64.
Wittwer, C.T., Herrmann, M.G., Moss, A.A. and Rasmussen, R.P. 1997. Continuous fluorescence monitoring of rapid cycle DNA amplification. Biotechniques 22: 130–139.
Xu, J., Xu, Y., Wang, H., Guo, C., Qiu, H., He, Y. and Meng, W. 2015. Occurrence of antibiotics and antibiotic resistance genes in a sewage treatment plant and its effluent-receiving river. Chemosphere 119: 1379–1385.
Yu, Y., Kim, J. and Hwang, S. 2006. Use of real-time PCR for group-specific quantification of aceticlastic methanogens in anaerobic processes: Population dynamics and community structures. Biotechnology and Bioengineering 93: 424–433.
Zarraonaindia, I., Smith, D.P. and Gilbert, J.A. 2013. Beyond the genome: Community-level analysis of the microbial world. Biology & Philosophy 28: 261–282.

5

In Vitro Diagnostics for Early Detection of Bacterial Wound Infection

Gregor Tegl,[1,*] *Andrea Heinzle,*[2] *Eva Sigl*[2] and *Georg M. Guebitz*[1]

1. Introduction

Bacterial contamination is the basis of wound infection and still constitutes a threat in health care. Despite significant developments of aseptic methods in the past century, a residual risk of bacterial contamination remains that pictures in about 10 per cent of postoperative wound infections (Petherick et al. 2006). A high risk of infection is also reported for chronic wounds (Sen et al. 2009), an issue predominantly affecting immune-suppressed people (Siddiqui and Bernstein 2010). As a consequence of manifestation of wound infection, wounds fail to heal. This greatly complicates the treatment modalities and life of the patients.

The infection progress strongly correlates with the interaction of invading bacteria with the wound environment. This process is divided into four sections, namely, contamination, colonization, critical colonization and infection, whereby impaired wound healing is considered to start at the stage of critical bacterial colonization (Schultz et al. 2003). To date applied detection methods are still based on the assessment of clinical signs, such

[1] Institute of Environmental Biotechnology, University of Natural Resources and Life Sciences Vienna, Konrad Lorenz Straße 20, 3430 Tulln an der Donau, Austria.
E-mail: guebitz@boku.ac.at
[2] Qualizyme Diagnostics GmbH & Co KG, Neue Stiftingtalstrasse 2, 8010 Graz, Austria.
E-mail: andrea.heinzle@qualizyme.com; eva.sigl@qualizyme.com
* Corresponding author: gregor.tegl@boku.ac.at

as redness (*rubor*), heat (*calor*), swelling (*tumor*), pain (*dolor*) and impairment of function (*functiolaesa*) which are strongly influenced by parameters, like impaired leukocyte functions of patients (Cutting and White 2005, Howell-Jones et al. 2005). These classical signs turn obvious at a progressed stage of infection and when microorganisms have already established a stable community, increasing resistance towards a variety of treatment strategies. Timely interference in the infection progression is of great importance and requires an effective online detection system indicating an infection at an early stage.

A variety of infection biomarkers is known based on microbes and their metabolites as well as based on intrinsic signal molecules secreted by the host immune system (Tegl et al. 2015). Since bacteria are the key players in infection, their early recognition is a reliable sign predicting an emerging infection. Several strategies are followed and described in the literature to directly assess bacteria concentrations in wounds but also indirectly detecting bacterial contamination by the aid of signal molecules. Available techniques in diagnostics enabled a substantial set of tools to be investigated for early stage detection of wound infection including synthetic materials, polysaccharides and proteins, amongst others (Fig. 1) (Schiffer et al. 2015).

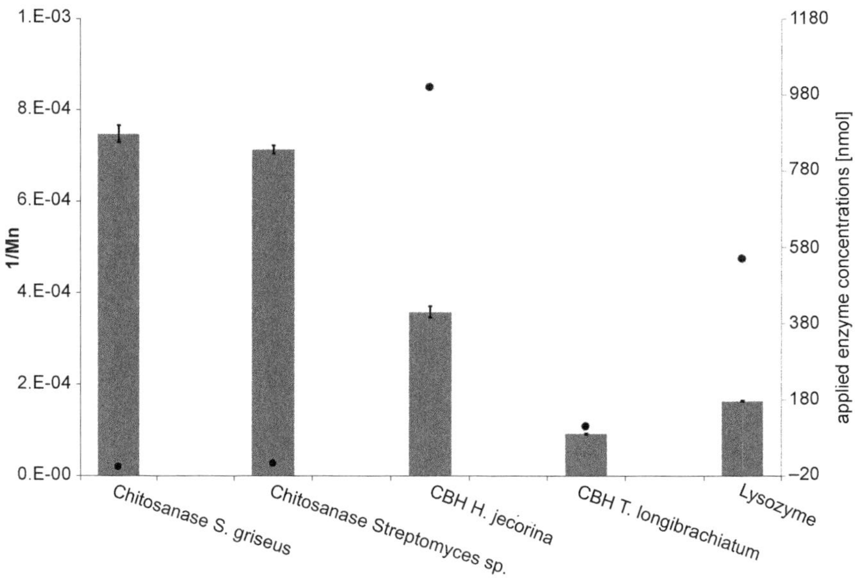

Figure 1. Enzyme-responsive polymers suitable for modification to detect wound infection. The different polymers are grouped into biopolymers (protein and polysaccharides) and synthetic polymers.

This chapter discusses the promising methods available for microbial infection detection divided in direct bacterial detection and indirect infection detection. The advantages and disadvantages of the available detection systems are elucidated and future perspectives in early detection of bacterial wound infection are discussed.

2. Direct Detection of Bacterial Wound Contaminants

The assessment of bacterial load in a wound represents a meaningful strategy since microbe concentration in a wound strongly correlates with the infection status (Bendy et al. 1964). Early research on microbes in wounds targeted an estimation of bacteria concentration that evokes wound infection, and the numbers always varied with the sampling technique chosen (Gardner et al. 2006). It is still questioned whether distinct bacterial species are responsible for an infection or if the emergence of a polymicrobial consortium is the crucial etiology (Bowler et al. 2001). However, bacteria are the key players in wound infection and thus constitute suitable biomarkers for infection detection. Initial detection methods were based on long-lasting analytical methods but substantial progress was made, resulting in novel strategies for a fast assessment of bacterial wound contamination (Tegl et al. 2015). The investigated methods thereby target either single bacterial species, but also consider the wound microbiome.

2.1 Near-real-time detection methods

Amongst the available detection systems, stimuli-responsive systems cover a broad pool of materials that change physical properties upon interaction with an external stimulus. Targeted modification of these materials can lead to real-time detection systems, enabling an application as point-of-care (PoC) diagnostics. An effective strategy was chosen by Zhou et al. producing lipid vesicles that were responsive towards virulence factors of pathogenic bacteria, like *S. aureus* and *P. aeruginosa* (Zhou et al. 2010). The working mechanism is based on bacteria-excreted toxins and lipases that also destroy human cell membranes and thus are capable of cleaving lipid-based vesicles. The respective vesicles, loaded with fluorescein or/and antimicrobial compounds, result in detection of pathogenic bacteria as well as in growth inhibition. The choice of lipids as responsive material further enables a discrimination between pathogenic and non-pathogenic bacteria, which commonly do not produce biological weapons against membranes. Subsequent refinements of the developed system could enhance stability of the lipid vesicles and functionalization on fabrics lead to a responsive system that proves effective for detecting pathogenic bacteria while not responding to *E. coli* (Zhou et al. 2011). In 2015, the vesicles were incorporated in a wound-dressing prototype and applied towards an *ex vivo* porcine skin burn wound model, finally proving the detection of pathogenic bacteria in a biofilm consortium (Fig. 2) (Thet et al. 2015). Detection of pathogens, *Enterococcus faecalis, P. aeuroginosa* and *S. aureus,* was realized after 6 hours-incubation time and thus counts as a suitable system for fast detection of wound infection. This system is of particular interest considering the bifunctional character of the vesicles as detection system as well as for antimicrobial treatment. Whether non-pathogenic bacteria have to be excluded or also contribute to an emerging infection remains part of the discussion.

Infection detection based on fluorescence and/or color signals depicts an efficient and easily assessable method. However, it requires the formulation of complex systems enabling a visualization process. Bioluminescence, by contrast, is a widely observed phenomenon induced by endogenous fluorescing molecules (Lee and Camilli 2000). Tissue components as well as a bunch of pathogenic bacteria, such as *S. aureus,* are auto-fluorescent and can thus be visualized. The spectral range of the respective fluorescence signals differs, enabling a distinction of connective tissue of a wound and contaminating bacteria (Moriwaki et al. 2011).

Wu et al. utilized the auto-fluorescence of both cell types to detect wound infection by the aid of auto fluorescent (AF) imaging (Wu et al. 2014).

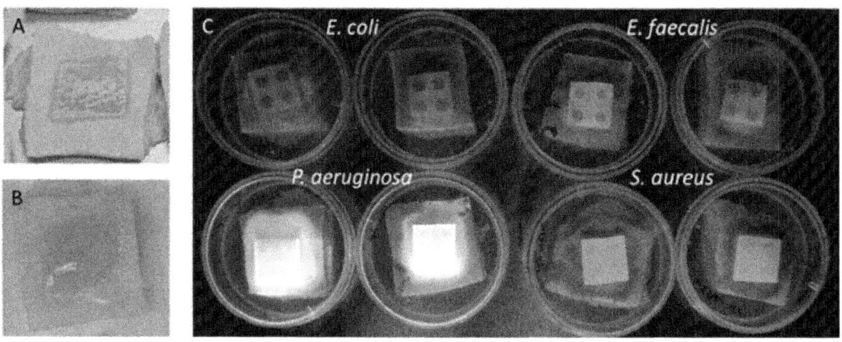

Figure 2. A wound dressing prototype based on bacteria-responsive fluorescent vesicles, applied on an infected porcine skin wound model. (A) a second degree partial thickness burn wound with blisters, (B) the burn wound 24 hours after infection with *P. aeruginosa,* and (C) the fluorescence response of prototype dressings on infected wounds of SPE (*Staphylococcus-Pseudomonas-Enterococcus*) group pathogens 24 hours later. The pathogenic bacteria *Enterococcus faecalis, P. aeuroginosa* and *S. aureus* showed a fluorescence response from the dressing, while the non-pathogenic reference *E. coli* did not show any signal. (Reproduced with permission from ACS Publications) (Thet et al. 2015).

White light (WL) and (AF) images were acquired using a charge-coupled device (CCD) camera whereby *S. aureus* could be detected via the red bioluminescence of endogenous porphyrins. Experiments were conducted to assess the mouse skin wound models inoculated with *S. aureus*, whereby a clear increase of bacteria concentration was visualized and also quantified to assess the spread area over time. Microbiological tests compared to AF imaging on chronic wounds of patients proved the suitability of the imaging system to detect multi-resistant *S. aureus* (MRSA). However, simultaneous detection of different strains is not possible with that system, but is of great importance due to significant variations in the polymicrobial flora within wounds that render single bacteria detection invalid (Gardner et al. 2001).

2.2 Analytical methods for infection detection

The need of PoC diagnostics for infection detection drives research to innovation in online measurement tools that provide fast information on the infection status. However, various methods were developed for bacteria detection in wounds that are based on analytical techniques, like mass spectrometry (MS) and microscopy (Tegl et al. 2015). The major disadvantage does not only arise from the duration of measurement and analysis, but much more from the withdrawal of wound samples and further preparation for analytics. Withdrawals of wound samples via swabbing are less reliable as they depend on the location of withdrawal, whereas quantitative biopsy constitutes the most accurate technique despite being invasive (Bill et al. 2001). Consequently, techniques relying on sample withdrawal are from many perspectives unfavorable, but possess analytical value and are thus considered in this chapter.

Capillary zone electrophoresis (CZE) could be utilized for selective detection of *E. coli* in wound fluids of infected wounds (Szeliga et al. 2011). Although a method not intended for whole cell analysis, a single peak was elucidated reliably to detect *E. coli* in wound samples in good sensitivity within a 30 min measurement. The advantage of the CZE method is the possibility to apply directly diluted wound fluid, which avoids complex sample preparation. *E. coli* can act as a pathogen in wound beds making detection meaningful. Nevertheless, the assessment of a single microbial species in a wound does not reliably indicate the infection status. Progress in microbe detection using CZE techniques could enable simultaneous detection of multiple cell types and was proposed by Rodriguez et al. (2006).

Beside capillary electrophoresis, mass spectrometry (MS) is another powerful tool that finds multiple applications in various research areas and routine analytics. Matrix-assisted laser desorption ionization time-of-flight (MALDI-TOF) mass spectrometry constitutes an efficient method for the analysis of complex biological samples and has gained considerable attention in identification of bacteria (Biswas and Rolain 2013). A standard procedure using Leuconostoc, Fructobacillus and Lactobacillus as reference species was implemented by De Bruyne et al. based on the comprehensive analysis using MS and machine learning (De Bruyne et al. 2011). This combination led to high accuracies in identification of the investigated bacteria. However, the approach was not applied on wound samples. Biswas and Rolain reviewed the capabilities and frontiers of MALDI-TOF MS for detection of aerobic and anaerobic bacteria, amongst others, and illustrated the advantages of MS. The crucial aspect in bacteria identification with MS is the restricted amount of reference spectra to date available, but required for bacteria detection. Expanding databases enable MALDI-TOF to act as fast and be less expensive first-line identification system, circumventing the immediate use of complex techniques like gene sequencing. Antibiotics treatment strategies could be fairly improved if knowledge is gained from MS reference databases, including multi-resistant bacteria. Despite the great potential of MS in infection detection, wound sample withdrawal and sample preparation still constitute a major burden.

Modern PCR and sequencing methods are without doubt the most accurate analytical techniques to detect multiple bacteria in complex samples (Melendez et al. 2010, Mazumdar et al. 2013). Simultaneous assessment of the entire wound microbiome could result in personalized treatment strategies. However, high instrumental costs and required sample withdrawals render these sophisticated approaches non-profitable for fast infection detection.

3. Indirect Detection Methods for Determination of Wound Infection

The core concept of online monitoring outlines early detection of an emerging infection and thus concentrates rather on infection prevention than on infection treatment. Assessment of the microbiome can unveil the pathogenomic picture of a wound, but constitutes an extrinsic biomarker not fully reflecting the activity of the host immune system. A variety of other infection biomarkers was found in infected wound beds, that give detailed information on bacteria concentration at an early phase of contamination. A substantial fraction of known biomarkers have also direct impact on the wound healing process, like the pH, which makes its detection particularly interesting. Biomolecules secreted in the wound environment either derive from bacterial metabolism, or can be a part of the human immune response.

3.1 Enzyme biomarkers for infection detection

The immune system tends to control biological processes by the aid of enzymes as it holds true for bacterial contamination of wounds. Activities of various enzymes, like myeloperoxidase and lysozyme, are down and up, regulated in response to the bacterial count and support the destruction of bacteria as well as damaged tissue (Schiffer et al. 2015, Sinclair and Ryan 1994). The infection status can thus be expressed as a function of enzyme activity and just requires transformation into a visual signal.

The so-far most intense elaborated detecting wound enzyme is based on modified polymer systems (Schiffer et al. 2015). The occurrence of distinct stimuli can be well visualized due to the change of physicochemical properties in response to external stimuli like enzymes. Amongst the variety of assays available to detect enzymes, color formation in response to enzyme activities is of particular interest considering the application in PoC diagnostics due to the ease in interpretation. Enzymes indicating wound infection include myeloperoxidase (MPO), lysozyme, human neutrophil elastase (HNE), cathepsin G and matrix metalloproteases (MMP). Their catalytic activity determines the polymer to be modified for detection of the respective enzyme, whereby infection-indicating enzymes are classified as hydrolases and oxidoreductases.

Proteolytic enzymes are excreted into the wound to destroy the damaged tissue. The respective responsive materials are thus mainly based on protein substrates (Heinzle et al. 2013). HNE was detected, using a peptide sequence (MeOSuc-AAPV) that was tagged with para-nitroaniline (pNA) and HNE-mediated hydrolysis, leading to release of yellow-colored pNA. The system was successfully tested to discriminate between infected and non-infected wound fluids with the free peptide as well as with immobilized derivatives on polyamides, polyester and protein-based polymers (Hasmann et al. 2011, Heumann et al. 2009, Fleck and Simman 2010). Non colorimetric detection methods are based on quartz crystal microbalance (QCM) measurements (Stair et al. 2009) and DNA aptamer sensors (He et al. 2012), but show restricted applicability for PoC devices as compared to color-changing detection systems.

Other elevated protease activities, like cathepsin G and MMPs, were successfully detected based on their ability to hydrolyze gelatin. Heinzle et al. produced gelatin beads that were loaded with Reactive Blue, leading to color release after incubation with gelatinolytic enzymes (Heinzle et al. 2013). This detection system exhibited optimal reaction kinetics for *ex vivo* diagnostics, visually discriminating infected wounds from non-infected wounds within 30 min. Direct detection of Cathepsin G was also proved via plasmon resonance imaging (SPRI) whereby a peptidyl inhibitor was immobilized on a gold chip and enabled quantification in high sensitivity and selectivity (Gorodkiewicz et al. 2012, Grzywa et al. 2014, Zou et al. 2012). The SPRI detection system further proved effective in measuring cathepsin G in complex samples like white blood cells, saliva samples and endometrial tissue.

Lysozyme is another important indicator of wound infection responsible for the cell wall degradation of wound-invading bacteria. It specifically attacks the peptidoglycan (PG) of cell walls hydrolyzing glycosidic linkages between N-acetylmuramic acid and N-acetylglucosamine residues that further results in cell destruction (Torsteinsdóttir et al. 1999, Salton 1957). Hydrolysis of PG dispersed in wound fluid leads to increasing transparency and visual detection was achieved by covalently attached Remazol Brilliant Blue. Colored oligosaccharides were released and led to significant absorbance differences in detecting elevated lysozyme activities in wound fluids (Hasmann et al. 2011). The lysozyme detection concept was extended and incorporated in a lateral flow system for the assessment of wound infection. The system is based on a size exclusion membrane that only allows migration of cleaved colored oligosaccharides, which are subsequently captured on a membrane (Schiffer et al. 2015). Intention to increase signal intensity led to the invention of enhanzymes that should multiply lysozyme-mediated PG hydrolysis by the aid of laccases (Schneider et al. 2012). Such a system could be incorporated in a lateral flow device that enables a fast and real-time assessment of the wound status. Alternative strategies for lysozyme detection are based on aptamer sensors and nanosensors. Both sensor systems show superior selectivity and sensitivity, but their application in complex samples like wound fluids is still restricted (Yoshida et al. 2014). Aptamer-based systems were successfully tested for lysozyme detection, including sensors using electrochemical impedance spectroscopy (Erdem et al. 2014, Rodríguez and Rivas 2009), fluorescent aptasensors (Chen et al. 2012, Rodríguez et al. 2009) and piezoelectric sensors (Lian et al. 2014). Nanosensors share advantages with aptamer sensors, like high sensitivity, but have also proven to work in complex media, such as in the case of a fluorescence nanosensor based on CdTe quantum dots (QDs) and carboxymethyl chitosan (CMCS) incorporating Zn^{2+} (Song et al. 2014). Lysozyme mediated hydrolysis of CMCS resulted in the release of Zn^{2+}, that led to quenching of photoluminescence and enabled quantitative lysozyme detection in human serum. A QCM nanosensor proved the same capability, but no information about measurements in human fluids is available yet (Sener et al. 2010).

MPO is another important biomarker, not only for wound infection, but also for several more disorders like cardiovascular diseases. In case of bacterial wound contamination, MPO activity is highly elevated, based on the increased production of neutrophils by the immune system (Hansson et al. 2006). It exhibits chlorination activity and thus produces hypochloric acid (HOCl), which is a bactericidal oxidative agent. Hasmann et al. could determine elevated MPO activity in wound fluids of infected wounds using guaiacol as substrate and MPO detection systems were rapidly extended to non-natural phenolic substrates (Hasmann et al. 2013). The phenol Fast Blue RR turned out MPO responsive and was functionalized with a siloxane spacer, enabling covalent immobilization on silica-based materials. Wound infection was determined by comparing MPO activity in infected and non-infected wound fluids and interference tests with hemoglobin further revealed the sensitivity of the detection system (Schiffer et al. 2015). The chlorination activity of MPO was utilized in detecting wound infection by the aid of an electrochemical sensor system (Hajnsek et al. 2015). A correlation of MPO oxidation and chlorination activity was visualized, using a hydrogen peroxide (H_2O_2) sensor that detected H_2O_2 consumption, which was supplied by using glucose oxidase, as a function of MPO activity. The produced biosensor could thereby clearly distinguish between wound samples from infected and non-infected wounds.

3.2 Metabolites and sensing molecules for infection detection

The variety of compounds contributing to the complexity of wound fluids are either secreted by the invading microflora or are part of the human immune response (Gethin 2012). Their concentrations nevertheless, reflects the wound infection status (Gethin et al. 2013). A majority of the biomarkers discussed in this subchapter are bacteria derived and also have impact on the wound-healing process.

An electrochemical impedance immune sensor was developed that combined the detection of both, immune system derived receptor-1 and MMP as well as N-3-oxo-dodecanoyl-1-homoserine lactone (HSL) that was unveiled as a quorum sensing molecule of *P. aeruginosa* (Ciani et al. 2012). Specific antibodies were immobilized on a gold solid phase electrolyte (SPE) and binding of the respective biomarkers resulted in an increase of electron transfer resistance. The applicability of the detection system was confirmed, measuring the mentioned analytes in mock wound fluids in good sensitivity.

Dealing with microbial infections, latest research focuses on quorum-sensing molecules of bacteria due to their crucial role in biofilm formation. Gene expression is up and down, regulated by the aid of sensing molecules and strongly effects bacterial virulence and thus the infection progress (Van Delden and Iglewski 1998). *P. aeruginosa* is a representative frequently found in infected wounds and is known to produce the redox active quorum sensor pyocyanin, whose concentration correlates with the bacterial load. Sharp et al. invented a sensor based on a carbon fiber tow electrode consisting of sandwiched carbon fibers electrically connected by a copper shielding tape that monitored pyocyanin via square wave voltammetry (Sharp et al. 2010). The sensor turned out selective, detecting *P. aeruginosa* in mixed bacteria populations. However, the proof of concept on wound fluid samples has not been published yet.

Another powerful wound infection indicator is the change of pH in the wound environment. The pH of an intact epidermis ranges from 4 to 6, while a triggered immune response was found to increase the pH, which also influences the wound healing process (Lars Schneider et al. 2007). Simple sensor systems for pH detection were prepared, based on common pH indicators, like bromocresol green and bromocresol purple that were immobilized on tetraethoxysilane films (Puchberger-Enengl et al. 2011). Chip-LED-mediated illumination of bromocresol green resulted in successful indication of pH changes in buffer solutions; the system was, however, inaccurate in artificial wound fluid. A pH detection system that proved its applicability in complex media was achieved, utilizing 2D luminescence imaging (Schreml et al. 2010). The use of fluorescein isothiocyanate (FITC) in combination with ruthenium (II)tris-(4,7-diphenyl-1,10-phenanthroline) ($Ru(dpp)_3$) and a time-gated CCD camera detected pH dependent luminescence. The pH responsive system was incorporated into microparticles that were further immobilized on polyurethane hydrogels to improve applicability.

A pH-changing compound accumulated in response to the host immune response in uric acid, which is produced by xanthine oxidase (XO) from purine derivatives (Fernandez et al. 2012). Disposable pH sensors for urate detection were produced by Phair et al. and were based on SPEs; they could successfully detect urate in buffer and blood samples (Phair et al. 2011). Further developments of carbon fiber sensors and pad-imprinted carbon-uric acid composite electrodes proved a sensitive detection of pH changes caused by uric acid in complex media, like artificial wound fluid, serum and blister fluid. However, these systems are lacking a definite proof on human wound fluid samples of infected wounds.

The majority of bacterial metabolites are small molecules, like ethanol, acetic acid and butyric acid (Bailey et al. 2008). A substantial percentage of these molecules are volatile (organic volatile compounds, VOC), thus detectable in the wound headspace, which circumvents sample withdrawals. Initial attempts to assess VOCs in the bacterial headspace were undertaken by Setkus et al. developing an SnO_2 gas sensor that was tested on cultures of *E. coli*, *P. aeruginosa* and *S. aureus* (Šetkus et al. 2006). Investigated compounds could be detected and were observed to vary with the growth stage of the bacteria. The great number of molecules led to the investigation of conducting organic polymer sensor arrays that were combined with neural network techniques (Bailey et al. 2008), but, important parameters like resistance and conductivity did not fulfill the requirements. Further improvements finally led to the invention of the e-nose consisting of a gas sensor array combined with a feature extraction method and a neural network classifier (Feng et al. 2011, Yan et al. 2012). The complex setup, including sampling unit, conditioning unit, and processing unit, was proven successful on single bacteria populations and VOCs could be detected in infected wounds of mice. Methods for background elimination and improvement of the feature extraction render this approach powerful for infection detection by the aid of an intelligent sensor system (He et al. 2012, Feng et al. 2013, Jia et al. 2014).

Keywords: Infection detection, Sensors, Diagnostics; Point of care, Chronic wounds, Polymers, Bacteria

References

Bailey, A., Pisanelli, A.M. and Persaud, K.C. 2008. Development of conducting polymer sensor arrays for wound monitoring. Sensors Actuators B Chem. 131(1): 5–9.

Bendy, R.H., Nuccio, P.A., Wolfe, E., Collings, B., Tamburro, C., Glass, W. and Martin, C.M. 1964. Relationship of quantitative wound bacterial counts to healing of decubiti: effect of topical gentamicin. Antimicrob. Agents Chemother. 10: 147–155.

Bill, T.J., Ratliff, C.R., Donovan, A.M., Knox, L.K., Morgan, R.F. and Rodeheaver, G.T. 2001. Quantitative swab culture versus tissue biopsy: A comparison in chronic wounds. Ostomy. Wound Manage. 47(1): 34–37.

Biswas, S. and Rolain, J.M. 2013. Use of MALDI-TOF mass spectrometry for identification of bacteria that are difficult to culture. J. Microbiol. Methods 92(1): 14–24.

Bowler, P.G., Duerden, B.I. and Armstrong, D.G. 2001. Wound microbiology and associated approaches to wound management. Clin. Microbiol. Rev. 14(2): 244–269.

Chen, C., Zhao, J., Jiang, J. and Yu, R. 2012. A novel exonuclease III-aided amplification assay for lysozyme based on graphene oxide platform. Talanta 101: 357–361.

Ciani, I., Schulze, H., Corrigan, D.K., Henihan, G., Giraud, G., Terry, J.G., Walton, A.J., Pethig, R., Ghazal, P., Crain, J., Campbell, C.J., Bachmann, T.T. and Mount, A.R. 2012. Development of immunosensors for direct detection of three-wound infection biomarkers at point of care using electrochemical impedance spectroscopy. Biosens. Bioelectron. 31(1): 413–418.

Cutting, K.F. and White, R.J. 2005. Criteria for identifying wound infection—revisited. Ostomy. Wound Manage. 51(1): 28–34.

De Bruyne, K., Slabbinck, B., Waegeman, W., Vauterin, P., De Baets, B. and Vandamme, P. 2011. Bacterial species identification from MALDI-TOF mass spectra through data analysis and machine learning. Syst. Appl. Microbiol. 34(1): 20–29.

Erdem, A., Eksin, E. and Muti, M. 2014. Chitosan-graphene oxide-based aptasensor for the impedimetric detection of lysozyme. Colloids Surf. B. Biointerfaces 115: 205–211.

Feng, J., Tian, F., Yan, J., He, Q., Shen, Y. and Pan, L. 2011. A background elimination method based on wavelet transform in wound infection detection by electronic nose. Sensors Actuators B Chem. 157(2): 395–400.

Feng, J., Tian, F., Jia, P., He, Q., Shen, Y. and Liu, T. 2013. Feature selection using support vector machines and independent component analysis for wound infection detection by electronic nose. Sensors Mater. 25(8): 527–538.

Fernandez, M.L., Upton, Z., Edwards, H., Finlayson, K. and Shooter, G.K. 2012. Elevated uric acid correlates with wound severity. Int. Wound J. 9(2): 139–149.

Fleck, C.A. and Simman, R. 2010. Modern collagen wound dressings: function and purpose. J. Am. Col. Certif. Wound Spec. 2(3): 50–54.

Gardner, S.E., Frantz, R.A., Saltzman, C.L., Hillis, S.L., Park, H. and Scherubel, M. 2006. Diagnostic validity of three swab techniques for identifying chronic wound infection. Wound Repair Regen. 14(5): 548–557.

Gardner, S.U.E.E., Frantz, R.A. and Doebbeling, B.N. 2001. The validity of the clinical signs and symptoms used to identify localized chronic wound infection. Wound Repair Regen. 9(3): 178–186.

Gethin, G. 2012. Understanding the inflammatory process in wound healing. Br. J. Community Nurs. Suppl. S17-S18: S20, S22.

Gorodkiewicz, E., Sieńczyk, M., Regulska, E., Grzywa, R., Pietrusewicz, E., Lesner, A. and Lukaszewski, Z. 2012. Surface plasmon resonance imaging biosensor for cathepsin G based on a potent inhibitor: development and applications. Anal. Biochem. 423(2): 218–223.

Grzywa, R., Gorodkiewicz, E., Burchacka, E., Lesner, A., Laudański, P., Lukaszewski, Z. and Sieńczyk, M. 2014. Determination of cathepsin G in endometrial tissue using a surface plasmon resonance imaging biosensor with tailored phosphonic inhibitor. Eur. J. Obstet. Gynecol. Reprod. Biol. 182C: 38–42.

Hajnsek, M., Schiffer, D., Harrich, D., Koller, D., Verient, V., Palen, J.V.D., Heinzle, A., Binder, B., Sigl, E., Sinner, F. and Guebitz, G.M. 2015. An electrochemical sensor for fast detection of wound infection based on myeloperoxidase activity. Sensors Actuators B Chem. 209(September): 265–274.

Hansson, M., Olsson, I. and Nauseef, W.M. 2006. Biosynthesis, processing, and sorting of human myeloperoxidase. Arch. Biochem. Biophys. 445(2): 214–224.

Hasmann, A., Gewessler, U., Hulla, E., Schneider, K.P., Binder, B., Francesko, A., Tzanov, T., Schintler, M., Van der Palen, J., Guebitz, G.M. and Wehrschuetz-Sigl, E. 2011. Sensor materials for the detection of human neutrophil elastase and cathepsin G activity in wound fluid. Exp. Dermatol. 20(6): 508–513.

Hasmann, A., Wehrschuetz-Sigl, E., Kanzler, G., Gewessler, U., Hulla, E., Schneider, K.P., Binder, B., Schintler, M. and Guebitz, G.M. 2011. Novel peptidoglycan-based diagnostic devices for detection of wound infection. Diagn. Microbiol. Infect. Dis. 71(1): 12–23.

Hasmann, A., Wehrschuetz-Sigl, E., Marold, A., Wiesbauer, H., Schoeftner, R., Gewessler, U., Kandelbauer, A., Schiffer, D., Schneider, K.P., Binder, B., Schintler, M. and Guebitz, G.M. 2013. Analysis of myeloperoxidase activity in wound fluids as a marker of infection. Ann. Clin. Biochem. 50(Pt 3): 245–254.

He, Q., Yan, J., Shen, Y., Bi, Y., Ye, G., Tian, F. and Wang, Z. 2012. Classification of electronic nose data in wound infection detection based on PSO-SVM combined with wavelet transform. Intell. Autom. Soft Comput. 18(7): 967–979.

Heinzle, A., Papen-Botterhuis, N.E., Schiffer, D., Schneider, K.P., Binder, B., Schintler, M., Haaksman, I.K., Lenting, H.B., Gübitz, G.M. and Sigl, E. 2013. Novel protease-based diagnostic devices for detection of wound infection. Wound Repair Regen. 21(3): 482–489.

Heumann, S., Eberl, A., Fischer-Colbrie, G., Pobeheim, H., Kaufmann, F., Ribitsch, D., Cavaco-Paulo, A. and Guebitz, G.M. 2009. A novel aryl acylamidase from nocardia farcinica hydrolyses polyamide. Biotechnol. Bioeng. 102(4): 1003–1011.

Howell-Jones, R.S., Wilson, M.J., Hill, K.E., Howard, A.J., Price, P.E. and Thomas, D.W. 2005. A review of the microbiology, antibiotic usage and resistance in chronic skin wounds. J. Antimicrob. Chemother. 55(2): 143–149.

Jia, P., Tian, F., He, Q., Fan, S., Liu, J. and Yang, S.X. 2014. Feature extraction of wound infection data for electronic nose based on a novel weighted KPCA. Sensors Actuators B Chem. 201: 555–566.

Lars Schneider, Andreas Korber and Stephan Grabbe, J.D. 2007. Influence of pH on wound-healing: A new perspective for wound-therapy? Arch. Dermatol. Res. 298(9): 413–420.

Lee, S. and Camilli, A. 2000. Novel approaches to monitor bacterial gene expression in infected tissue and host. Curr. Opin. Microbiol. 3(1): 97–101.

Lian, Y., He, F., Mi, X., Tong, F. and Shi, X. 2014. Lysozyme aptamer biosensor-based on electron transfer from SWCNTs to SPQC-IDE. Sensors Actuators B Chem. 199: 377–383.

Mazumdar, R.M., Chowdhury, A., Hossain, N., Mahajan, S. and Islam, S. 2013. Pyrosequencing—a next generation sequencing technology. World Appl. Sci. J. 24(12): 1558–1571.

Melendez, J.H., Frankel, Y.M., An, A.T., Williams, L., Price, L.B., Wang, N.-Y., Lazarus, G.S. and Zenilman, J.M. 2010. Real-time PCR assays compared to culture-based approaches for identification of aerobic bacteria in chronic wounds. Clin. Microbiol. Infect. 16(12): 1762–1769.

Moriwaki, Y., Caaveiro, J.M.M., Tanaka, Y., Tsutsumi, H., Hamachi, I. and Tsumoto, K. 2011. Molecular basis of recognition of antibacterial porphyrins by heme-transporter IsdH-NEAT3 of *Staphylococcus aureus*. Biochemistry 50: 7311–7320.

Petherick, E.S., Dalton, J.E., Moore, P.J. and Cullum, N. 2006. Methods for identifying surgical wound infection after discharge from hospital: A systematic review. BMC Infect. Dis. 6: 170.

Phair, J., Newton, L., McCormac, C., Cardosi, M.F., Leslie, R. and Davis, J. 2011. A disposable sensor for point of care wound pH monitoring. Analyst 136(22): 4692–4695.

Puchberger-Enengl, D., Krutzler, C., Vellekoop, M.J. and Gusshausstrasse, E. 2011. Organically modified silicate film pH sensor for continuous wound monitoring. In Sensors, IEEE, Vol. 11679, 682: 679–682.

Rodriguez, M.A., Lantz, A.W. and Armstrong, D.W. 2006. Capillary electrophoretic method for the detection of bacterial contamination. Anal. Chem. 78(14): 4759–4767.

Rodríguez, M.C. and Rivas, G.A. 2009. Label-free electrochemical aptasensor for the detection of lysozyme. Talanta 78(1): 212–216.

Salton, M.R. 1957. The properties of lysozyme and its action on microorganisms. Bacteriol. Rev. 21(2): 82–100.

Schiffer, D., Blokhuis-Arkes, M., van der Palen, J., Sigl, E., Heinzle, A. and Guebitz, G.M. 2015. Assessment of infection in chronic wounds based on the monitoring of elastase, lysozyme and myeloperoxidase activities. Br. J. Dermatol.

Schiffer, D., Tegl, G., Heinzle, A., Sigl, E., Metcalf, D., Bowler, P., Burnet, M. and Guebitz, G.M. 2015. Enzyme-responsive polymers for microbial infection detection. Expert Rev. Mol. Diagn. 15(09): 1–7.

Schiffer, D., Tegl, G., Vielnascher, R., Weber, H., Schoeftner, R., Wiesbauer, H., Sigl, E., Heinzle, A. and Guebitz, G. 2015. Fast blue RR—siloxane derivatized materials indicate wound infection due to a deep blue color development. Materials (Basel). 8(10): 6633–6639.

Schiffer, D., Verient, V., Luschnig, D., Blokhuis-Arkes, M.H.E., Palen, J.V.D., Gamerith, C., Burnet, M., Sigl, E., Heinzle, A. and Guebitz, G.M. 2015. Lysozyme-responsive polymer systems for detection of infection. Eng. Life Sci. 15: 368–375.

Schneider, K.P., Gewessler, U., Flock, T., Heinzle, A., Schenk, V., Kaufmann, F., Sigl, E. and Guebitz, G.M. 2012. Signal enhancement in polysaccharide based sensors for infections by incorporation of chemically modified laccase. N. Biotechnol. 29(4): 502–509.

Schreml, S., Meier, R.J., Wolfbeis, O.S., Landthaler, M. and Szeimies, R. 2010. 2D Luminescence Imaging of pH *in vivo*. PNAS.

Schultz, G.S., Sibbald, R.G., Falanga, V., Ayello, E.A., Dowsett, C., Harding, K., Romanelli, M., Stacey, M.C., Teot, L. and Vanscheidt, W. 2003. Wound bed preparation: A systematic approach to wound management. Wound Repair Regen. 11(s1): S1–S28.

Sen, C.K., Gordillo, G.M., Roy, S., Kirsner, R., Lambert, L., Hunt, T.K., Gottrup, F., Gurtner, G.C. and Longaker, M.T. 2009. Human skin wounds: A major and snowballing threat to public health and the economy. Wound Repair Regen. 17(6): 763–771.

Sener, G., Ozgur, E., Yılmaz, E., Uzun, L., Say, R. and Denizli, A. 2010. Quartz crystal microbalance-based nanosensor for lysozyme detection with lysozyme imprinted nanoparticles. Biosens. Bioelectron. 26(2): 815–821.

Šetkus, A., Galdikas, A.-J., Kancleris, Ž.-A., Olekas, A., Senulienė, D., Strazdienė, V., Rimdeika, R. and Bagdonas, R. 2006. Featuring of bacterial contamination of wounds by dynamic response of SnO_2 gas sensor array. Sensors Actuators B Chem. 115(1): 412–420.

Sharp, D., Gladstone, P., Smith, R.B., Forsythe, S. and Davis, J. 2010. Approaching intelligent infection diagnostics: carbon fibre sensor for electrochemical pyocyanin detection. Bioelectrochemistry 77(2): 114–119.

Siddiqui, A.R. and Bernstein, J.M. 2010. Chronic wound infection: facts and controversies. Clin. Dermatol. 28(5): 519–526.

Sinclair, R.D. and Ryan, T.J. 1994. Proteolytic enzymes in wound healing: the role of enzymatic debridement. Australas. J. Dermatol. 35(1): 35–41.

Song, Y., Li, Y., Liu, Z., Liu, L., Wang, X., Su, X. and Ma, Q. 2014. A novel ultrasensitive carboxymethyl chitosan-quantum dot-based fluorescence "turn on-off" nanosensor for lysozyme detection. Biosens. Bioelectron. 61: 9–13.

Stair, J.L., Watkinson, M. and Krause, S. 2009. Sensor materials for the detection of proteases. Biosens. Bioelectron. 24(7): 2113–2118.

Szeliga, J., Kłodzińska, E. and Jackowski, M. 2011. The clinical use of a fast screening test based on technology of capillary zone electrophoresis (CZE) for identification of *Escherichia coli* infection in biological material. Med. Sci. Monit. 17(10): 91–96.

Tegl, G., Schiffer, D., Sigl, E., Heinzle, A. and Guebitz, G.M. 2015. Biomarkers for infection: enzymes, microbes, and metabolites. Appl. Microbiol. Biotechnol. 99(11): 4595–4614.

Thet, N.T., Alves, D.R., Bean, J.E., Booth, S., Nzakizwanayo, J., Young, A.E.R., Jones, B.V. and Jenkins, A.T.A. 2015. Prototype development of the intelligent hydrogel wound dressing and its efficacy in the detection of model pathogenic wound biofilms. ACS Appl. Mater. Interfaces, 151022154036007.

Torsteinsdóttir, I., Hâkansson, L., Hällgren, R., Gudbjörnsson, B., Arvidson, N.G. and Venge, P. 1999. Serum lysozyme: A potential marker of monocyte/macrophage activity in rheumatoid arthritis. Rheumatology (Oxford) 38: 1249–1254.

Van Delden, C. and Iglewski, B.H. 1998. Cell-to-cell signaling and *Pseudomonas aeruginosa* infections. Emerg. Infect. Dis. 4(4): 551–560.

Wu, Y.C., Kulbatski, I., Medeiros, P.J., Maeda, A., Bu, J., Xu, L., Chen, Y. and DaCosta, R.S. 2014. Autofluorescence imaging device for real-time detection and tracking of pathogenic bacteria in a mouse skin wound model: preclinical feasibility studies. J. Biomed. Opt. 19(8): 085002.

Yan, J., Tian, F., He, Q. and Shen, Y. 2012. Feature extraction from sensor data for detection of wound pathogen based on electronic nose. Sensors Mater. 24(2): 57–73.

Yoshida, W., Abe, K. and Ikebukuro, K. 2014. Emerging techniques employed in aptamer-based diagnostic tests. Expert Rev. Mol. Diagn. 14(2): 143–151.

Zhou, J., Loftus, A.L., Mulley, G. and Jenkins, A.T.A. 2010. A thin film detection/response system for pathogenic bacteria. J. Am. Chem. Soc. 132(18): 6566–6570.

Zhou, J., Tun, T.N., Hong, S., Mercer-Chalmers, J.D., Laabei, M., Young, A.E.R. and Jenkins, A.T.A. 2011. Development of a prototype wound dressing technology which can detect and report colonization by pathogenic bacteria. Biosens. Bioelectron. 30(1): 67–72.

Zou, F., Schmon, M., Sienczyk, M., Grzywa, R., Palesch, D., Boehm, B.O., Sun, Z.L., Watts, C., Schirmbeck, R. and Burster, T. 2012. Application of a novel highly sensitive activity-based probe for detection of cathepsin G. Anal. Biochem. 421(2): 667–672.

6

Biofilm Impedance Monitoring

Jacobo Paredes, Imanol Tubía* and *Sergio Arana*

1. Introduction

Biofilms are 3D **complex microbial communities** composed of bacterial cells irreversibly associated and attached to a surface, and encapsulated in their own extracellular matrix (Donlan 2002). These are highly heterogeneous biological systems that can be formed by one or more species of microorganisms (Stewart and Franklin 2008). The most relevant characteristic of these formations is their **high resistance to antimicrobial treatment**. It has been reported that necessary concentrations to kill the microorganisms can be between 500 and 5000 times higher than those for non-biofilm bacteria (del Pozo et al. 2009). The Extracellular Matrix (ECM) is mainly composed by polysaccharides, some lipids and proteins and water that accounts for between 75 and 90 per cent of the biofilm (Costerton et al. 1999, Sabater et al. 2007). Within the **ECM** there is a gradient of chemical concentration, nutrients, pH, etc. that cause bacteria to adapt to these environments, change their growth kinetics and even their genotype and phenotype (Davies 2003). Also, bacteria within the biofilm are able to communicate through different molecular mechanisms known as ***quorum sensing*** that provides even more different behavior than planktonic cells (Donlan and Costerton 2002). The direct consequence of these arrangements is extreme difficulty to kill all bacteria with traditional treatments.

CEIT and Tecnun (University of Navarra), Paseo de Manuel Lardizábal, n° 15, 20018 Donostia-San Sebastián, Spain.
* Corresponding author: jparedes@ceit.es

Biofilm colonization is a common problem along in many different environments due to the **ubiquity of microorganisms** and their low-demanding growth conditions. There are a number of areas of interest that are affected by these problems, ranging from healthcare environment (Coenye and Nelis 2010, Harris and Richards 2006, Williams and Bloebaum 2010), food industry (Ivnitski et al. 2000, Silley and Forsythe 1996, Yang and Bashir 2008), merchant vessels hull (Blenkinsopp and Costerton 1991, Coenye and Nelis 2010) to water or oil pipelines (Blenkinsopp and Costerton 1991, Muñoz-Berbel et al. 2006), etc., generating numerous problems with important economic losses. Figure 1 shows a representation of the biofilm development process by a step by step scheme. Also there are represented the most important problematic areas in general fields as discussed previously.

There are certain conditions regarding flow dynamics, anchoring sites, nutrients and media characteristics, under which **biofilm formation and development is favored**. Steady conditions (pH, nutrient concentration, flow) would promote quick attachment and growth of bacteria-biofilm formations over inert surfaces. On the other hand, it has been reported that some shear stress can promote biofilm development through nutrients renewal (Donlan and Costerton 2002), as well as dispersal and spreading of biofilm to surrounding areas (Hall-Stoodley et al. 2004, McLandsborough et al. 2006). Therefore, sites or locations like edges, corners, changes of direction will be predetermined to facilitate biofilm colonization.

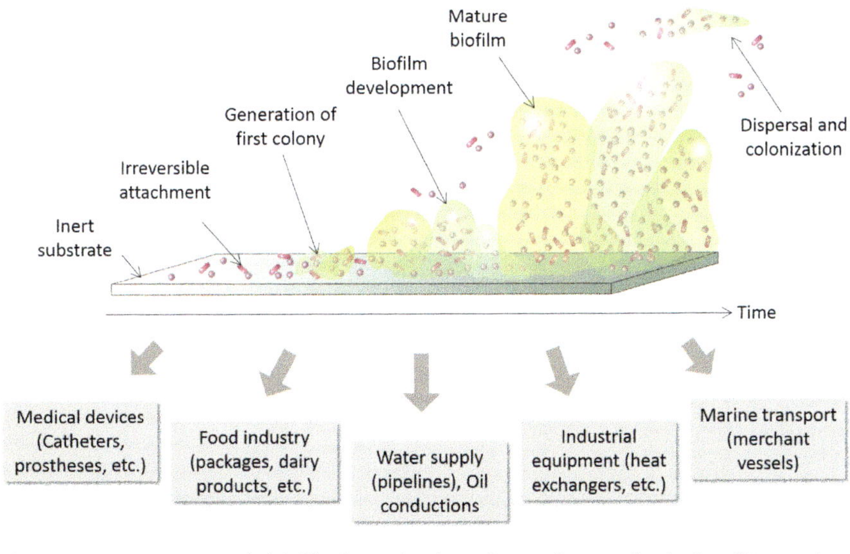

Figure 1. Representation of a biofilm formation from the attachment of a single cell to a mature biofilm and the potential problematic and monitoring need.

Efforts are focused on developing **preventing strategies** against bacterial attachment and therefore, biofilm formation. Numerous works have been published in this direction and the most common methods are based on material coatings, nanoparticles, polymers, etc. or preventing attachment with hydrophobic surfaces and materials (Knetsch and Koole 2011, Muñoz-Bonilla and Fernández-García 2012, Siedenbiedel and Tiller 2012).

Also in the literature, it is possible to find **alternative treatment approaches** that are essentially based on weakening the extracellular matrix or disturbing bacterial communication (i.e., against Quorum sensing) (Stewart 2003). These methods are to be combined with traditional antibiotic treatments with increasing efficiency by promoting drug diffusion in the biofilm structure. Other innovative strategies include looking for ways to prevent the initial attachment as mentioned before. And other approaches try to take advantage of electric fields to direct the drugs inside the ECM of the biofilm (del Pozo et al. 2009, Giladi et al. 2008, Sultana et al. 2015).

In this context, **early detection of bacterial biofilm** formations has been postulated as one effective approach to prevent further development of bacteria and perform efficient treatments. Antimicrobial treatments will provide a better outcome when acting against immature biofilms and planktonic cells. An *in situ* detection approach is also key for effectively monitoring the presence or development of biofilm in real time. Therefore, simple and effective tools would provide real time and *in situ* analysis on the target site allowing efficient prophylactic actions. In this sense, a good example of the latest trends might include biofilm monitoring and bacterial detection using smartphones (Jiang et al. 2014) wireless and battery-less biosensors, etc.

Impedance microbiology has been proposed as a suitable technique that allows monitoring of bacterial biofilms growth in real time. Since Cady et al. described the working principles of this method (Cady et al. 1978a, 1978b), it has been largely applied in microbiology detection for different fields: food borne pathogen detection (Yang and Bashir 2008), industrial biocorrosion monitoring (Dheilly et al. 2008), biochip analysis (Gomez et al. 2001), clinical sample analysis (Oliver et al. 2006, Paredes et al. 2012), etc.

This technique can be performed under a vast range of different environments. It is clear that impedance analysis is an **easy-to-integrate technique** that can be adapted by the electrodes or microelectrodes design to any geometrical requirements. Moreover, impedance is usually analyzed in the range of the frequency to identify the most sensitive parameters. Many authors also agree that multiparametric measurements will provide more accurate and complete information about biofilm development even when the other parameter changes could be considered negligible; parameters, such as pH, ion concentration, current flow, electrochemical techniques, etc. (Estrada-Leypon et al. 2015).

The aim of this chapter is to **review the state-of-the-art** methods for impedimetric detection of biofilms and their monitorization. Over the last years there are many new investigations on this subject using different technologies. A comparison of the most relevant works in this field is also discussed in order to highlight the most interesting advances and methodologies. Most importantly, a comparison between direct detection methods for bacterial cells and biofilm development monitoring will be discussed. Conclusions drawn from this analysis would help the development and implementation of new *in situ* detection systems improving quality, efficiency and security in many different environments.

2. Impedance Microbiology

2.1 Working principle

Electric impedance is defined as the relationship between the applied electric field and the electric current in the complex domain (Y. Wang et al. 2012). Every material presents a unique impedance characteristic, which is the opposition of the material to the current flux under an alternative electric field. Equation (1) shows the Ohm's law in the complex field expressing this relationship, and equation (2) expresses the impedance as the sum of its terms, resistance and reactance.

$$\vec{Z}\,(w) = \vec{V}(t)/\vec{I}(t) \tag{1}$$

$$\vec{Z} = Z' + jZ'' = R + jX = |Z|\,\cos\varphi + j|Z|\,\sin\varphi \tag{2}$$

where Z is the impedance and Z′ and Z″ its two components real (the resistance: R) and imaginary (the reactance: X), respectively. φ is the phase of the impedance that shows its capacitive or inductive character.

Impedance depends on many factors mainly from the material (solid or liquid) but also on the measurement setup and parameters: **frequency, applied voltage, connections, temperature, etc**. The most common frequency range in impedance spectroscopy for bacterial monitoring varies from hundreds of **mHz to MHz**. Typical values for the electric field are in the order of mV and everything is performed under the real temperature of the application (should be fairly constant).

One can also distinguish between two types of measurements depending on the interaction of the sample with the electrodes: non-faradaic measurements, also known as **dielectric impedance**, in which there is no electronic transfer between sample and electrode but rather charge movements (ionic species orientation, adsorption phenomena, etc.); and faradaic measurements which involve **electrochemical** reactions (redox) (Varshney and Li 2009). Obviously the measurement setup and parameter values for each measurement will be different. Bacterial biofilms can be measured by both techniques, depending on the sample electrochemical characteristics. Usually impedance variations are analyzed by the representation of the relative changes from a given baseline.

Biofilm monitoring is based on impedance changes that in general vary according to bacterial concentration and biofilm development (Gómez-sjöberg et al. 2005, Yang and Bashir 2008). There are two main effects that cause these variations: **bacterial metabolism and biomass increase** (biofilm growth). Both effects will change how the electric current flows from one electrode to the other. On the one hand, bacterial metabolism consumes oxygen and nutrients and produces carbon dioxide and organic acids (Gomez et al. 2002); on the other, the increasing number of bacterial cells affects as well the conductivity. The cellular membrane (bilipidic layer) facilitates ion transfer through specific ionic channels. Moreover the cellular membrane act as electrical capacitor storing charge under alternative electric fields.

Figure 2 represents the three main configurations of electrodes that have been used for biofilm monitoring. These microbes create biofilms on every free surface including the electrodes as it is represented in the figure. Changes in conductivity are therefore caused by the coating of the electrodes by the formed biofilm and the changes in ionic concentration due to the

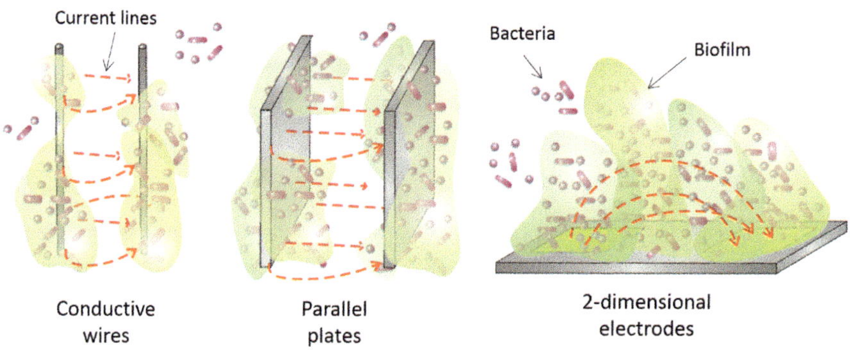

Figure 2. Schematic of the working principle of impedance measurements and monitoring using wires, parallel plates and 2D configurations. A mature biofilm was depicted as it was growing attached to those surfaces.

microbes' metabolism. Depending on the flow dynamics, the location of the electrodes and other parameters it is possible to determine which effect is predominant in the impedance change.

The first case of the figure presents two wires (usually platinum, but could be as well stainless steel and other metals) submerged within the microbial culture; the second case presents two bulk parallel macro electrodes submerged in the target sample containing microorganisms. Even though the drawing shows a large biofilm over the electrodes, this might be exaggerated from the reality, but with enough time and nutrients its biofilm will definitely colonize everything around. And finally in the third depicted case there are surface electrodes (2-dimensional) over a non-conductive substrate. Biofilm development over the electrodes will change the conductivity across them. The same principle is applied in this case.

Also the figure contains a representation of the electric current lines that are generated when the electric field is applied across them. Depending on what those current lines encounter in their way from one electrode to another, will be the impedance values.

2.1.1 Origins of impedance microbiology

Monitoring microbial growth was firstly described by Steward in 1899 (Stewart 1899). Other works followed during those early days, researchers like Ur and Brown (Ur and Brown 1975), or Cady et al. (Cady et al. 1978a, 1978b). These original investigations led to the development of automated equipment, such as the bactometer® or the Malthus® systems. Nowadays, there are numerous instruments for routine analysis of different kinds of samples.

All these first commercial instruments plus two others: RABIT® or Bac-Trac®, are based on evaluating the metabolic activity trend in a known media (Y. Wang et al. 2012). Samples are placed in the automated equipment for monitoring impedimetric (or conductimetric) changes over a period of time. Infected samples are sorted out based on thresholds of relative changes of different tests.

Figure 3 shows some of the first setups these authors used for bacterial growth monitoring by means of conductimetric changes. All setups were designed for small volumes of samples using different electrode configurations such as, parallel plates, two dimensional printed electrodes or cylindrical electrodes inside a capillary tube.

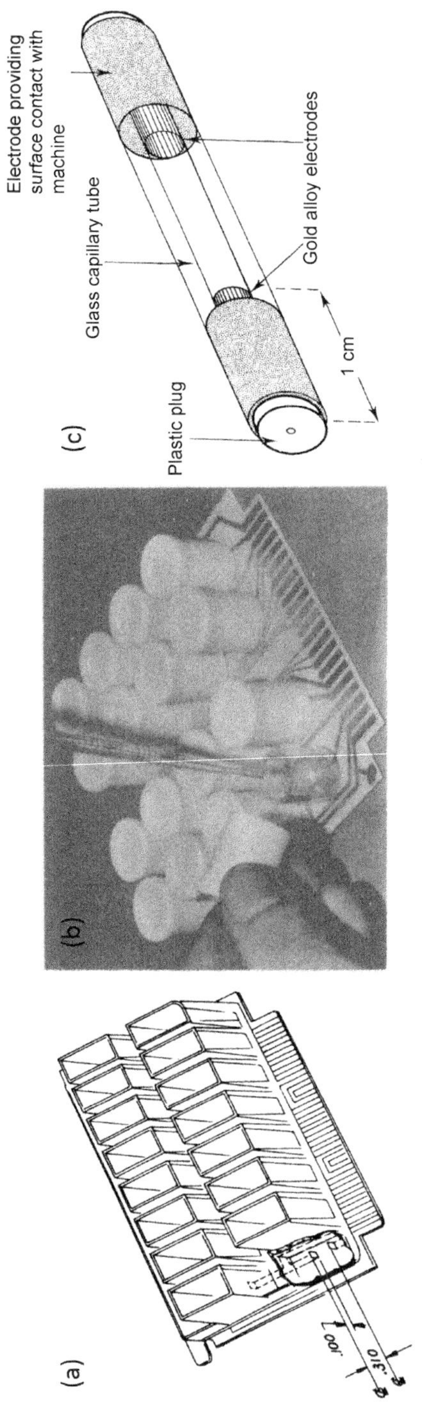

Figure 3. (a) Drawing of the module for incubation and impedance measurement with stainless steel macro electrodes developed by Cady et al. (b) Photograph of a printed board with bonded wells for samples from Cady et al. (c) Impedance monitoring cell developed by Ur and Brown.

2.2 Culturing setup design and configuration

Growth kinetics are strongly affected by the environmental conditions, nutrients, dissolved gases, temperature, flow dynamics, etc. Therefore, it is extremely important to **know the environment** and how all factors affect the growth rate of the biofilm in each specific scenario.

Especially for *in situ* biofilm control through impedance monitoring, this knowledge comes to be a critical point. This statement is also valid for *in vitro* experimentation, where researchers try to **simulate and mimic the real conditions** of a specific application to develop new monitoring tools. In fact, so many different setups have been proposed in the literature and sometimes it is difficult to compare them or understand how they work.

Figure 4 gathers a representation of the most common culture setups for biofilm research by means of impedance measurements: Medium size volumes, such as a reactor systems (i.e., Centers for Disease Control, CDC Biofilm Reactor, Biosurface Technologies Corp, Boezemann, MT), plates and multi-well plates, micro chambers or micro-wells on chips and microfluidic devices. The purpose of these devices is the detection and monitoring of biofilm development. However they can be also used for non-forming biofilm bacterial detection.

All these 5 devices represent a wider variety of other setups found in bibliography that could be included in at least one of the groups. Other setups might present variations of parameters like volume, materials, number of inlets, number of sensors, agitation conditions, etc.

Table 1 shows a comparison of the most important characteristics of the previously described setups such as: volume, dynamic conditions, flexibility, combination with other analytical techniques (microscopy), etc. All the culturing supports are required to be sterilized and to maintain the optimum culturing conditions: temperature and humidity.

Given this analysis, it is clear that depending of the experimental purpose there are more adequate supports for experimentation than others. It is important to mention that these setups are a simplistic representation of the real environment; for example, a pipeline can be hardly represented by a microfluidic device but it will certainly provide interesting results that might help better understand the problem. There are other environments that will certainly be represented more faithfully—food packages, beverages, small volumes stagnant liquids, containers, etc.

According to the previous table, setups can be classified into different groups: according to the dynamic conditions, if they allow sampling during the experimentation or simultaneous visualization in a microscope, etc. These characteristics are extremely important because usually it is necessary

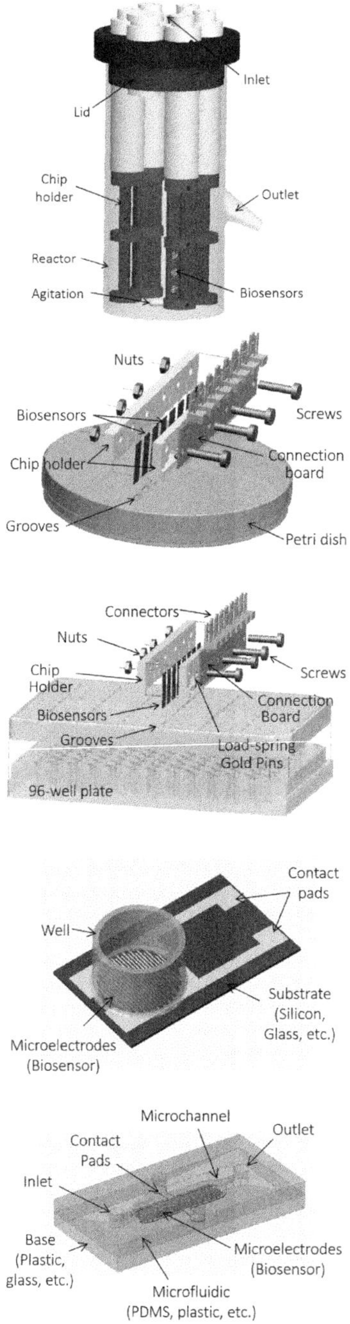

MODIFIED CDC REACTOR

Culturing reactors consist of a vessel with capacity for several hundreds of mili liters. Biosensors are integrated in direct contact with the microbial culture. This device can be used as a small model of bigger containers, mimicking the dynamic conditions. With flexible operation these device can be used for many different applications. A CDC reactor system is one of the standards setups for biofilm culture. It is broadly used for biofilm research.

References: (J. Paredes et al., 2014; Paredes et al., 2012)

CULTURE DISHES

This type of devices integrates a number of biosensors in standard culturing dishes (i.e. petri). So it is possible to take advantage of all previous knowledge and control the growth conditions. It is an effective and affordable method to monitor biofilm performance using a set of different biosensors. However these setups are not flexible for dynamic culture conditions.

References: (J. Paredes et al., 2014; Jacobo Paredes et al., 2014)

WELL PLATES

Plastic culture plates are commonly used in biofilm research, and therefore there is a vast knowledge about their performance. It is possible to integrate biosensors within these systems to monitor different parameters of biofilm growth. Its main advantage is the high number of samples allowed in just one set of experiments. Although these are always static conditions.

References: (J. Paredes et al., 2014; Paredes et al., 2013)

MICROWELL ON CHIPS

These systems are composed of a substrate containing the microelectrodes (biosensor) and a well bonded/placed on top. These devices allowed a complete design of the experimental setup: flow conditions (for closed chambers), number of integrated sensors, easy access to the connections, materials, etc. However these devices trend to be more expensive and complex to handle.

References: (J Paredes et al., 2014; J. Paredes et al., 2014)

MICROFLUIDIC DEVICES

Microfluidic devices present a unique potential for biofilm research under a thorough controlled environment. They present a flexible design to achieve different purposes and allow the use of a number of materials. It is also possible to modify on real time the flow conditions, media and other parameters.

References: (Kim et al., 2015; Varshney et al., 2007)

Figure 4. Representation of the most common setups used for bacterial biofilm culturing and monitoring (Paredes et al. 2014).

Table 1. Comparison between microelectrode design and methods for biofilm detection.

Setup	Vol.	Bacteria	Media	(μm)		Volt.	Freq.	Materials		Detection limit	Max. variation			Ref.
				Width	Gap			Electrode	Substrate		Param.	Freq.	Time	
Immobilized antibodies														
Cuvette	100 μl	*S. typhimurium*	Food sample	10	2, 5, 10	50 mV	100-1 MHz	Cr + Au	Glass	10^3	Z	100 Hz	10 min	(Kim et al. 2008)
Cuvette	20 mL	*E. coli*	Peptone water	3	4	50 mV	10-13 MHz	Ti + Au	SiO_2 (2 μm)	10^4	Z	1 kHz	5 min	(Radke et al. 2004)
Cuvette	30 mL	*E. coli* *S. infantis*	Peptone water	3	4	50 mV	100-10 MHz	Ti + Au	SiO_2 (2 μm)	10^4	Z	1 kHz	10 min	(Radke and Alocilja 2005)
Mounted chamber	100 μL	*S. aureus* *S. epidedrmidis*	Luria Broth	2, 5 10	2	-	1k-1 MHz	Al + Al_2O_3	SiO_2 (350 nm)	10^7	C	1 kHz	10 min	(Tang et al. 2011)
Drop	0,5 μL	*E. coli*	DI water	4	2	0,5V	DC	Ti + Au	SiO_2 (0,5 μm)	10^0	ρ	-	~s	(Lu et al. 2008)
Antibodies and secondary antibody														
Drop	-	*E. coli*	BHI	200	200	10 mV	1 Hz-1 MHz	Au	-	10^2	EIS	100 Hz	1 h	(Li et al. 2015)
Antimicrobial peptides and other molecules														
Mounted chamber	~ mL	*E. coli*	PBS and buffers	100	40	-	20-160 Hz	Au	Glass	10^2	Z	40 Hz	500 s	(Jiang et al. 2015)
Conjugated Magnetic nanoparticles and antibodies														
Microfluidic chamber	25 μL	*E. coli* *S. typhimurium*	DI wáter and PBS	15	15	50 mV	1-100 kHz	Au	Glass	10^4	Z	1 kHz	~s	Yang 2008 (Lu et al. 2008)
Microfluidic chamber	2 μL	*E. coli*	Meat solution	15	15	100 mV	10-1 MHz	Cr + Au	Glass	10^4	Z	40 kHz	35 min	(Varshney and Li 2007)
Drop	2 μL	*E. coli*	Meat solution	15	15	100 mV	10-1 MHz	Cr + Au	Glass	10^4	Z	16 kHz	35 min	(Varshney et al. 2007)
Mounted chamber	200 μL	*E. coly* *S. typhimurium*	Meat solution	200	200	5 mV	10-1 MHz	Au	PCB	10^3	EIS	1 kHz	30 min	(Xu et al. 2016)

Table 1 contd. ...

...Table 1 contd.

Setup	Vol.	Bacteria	Media	(μm)		Volt.	Freq.	Materials		Detection limit	Max. variation			Ref.
				Width	Gap			Electrode	Substrate		Param.	Freq.	Time	
Dielectrophoresis effect (Concentration/trapping)														
Microfluidic chamber	μL/h flow	*E. coli*	Manitol	50	5	5 V	100 kHz	Cr	Glass	-	ρ	1 kHz	-	(Hamada et al. 2013)
Mounted chamber	200 μL	*E. coli*	Manitol	50	5	-	100 kHz, 1 MHz	-	-	10^5	ρ	1 MHz	-	(Suehiro et al. 2003)
Microfluidic chamber	1500 μL/h	*E. coli*	Water	20	5	2.5 V	1 kHz-10 MHz	Cr + Au	Glass	3×10^2	ρ	100 kHz	1 min	(Kim et al. 2015)
Microfluidic chamber	1 μL	*S. epidermidis*	PBS	2	4	50 mV	1 kHz-10 MHz	Al + Al_2O_3	Glass	3×10^2	ρ	-	20 min	(Couniot et al. 2015)
Surface-free transducer														
Microfluidic chamber	(33 μl/min)	*E. coli*	YPLT	15	15	100 mV	10-1 MHz	Ti-Tu + Au	Glass	10^2	Z	10 Hz	14 h	(Varshney and Li 2008)
Cuvette	4,5 mL	*S. typhimurium*	Milk sample	15	15	5 mV	0,2-5 MHz	ITO + Ti	Glass	10^4	Z	10 Hz	10	(Yang et al. 2004)
Microfluidic chamber	5,3 nL	*L. monocytogenes* *E. coli*	Low cond. media	50	80	50 mV	100-1 MHz	Pt	SiO_2 (0,45 μm)	10^5	Z		~s	(Gomez et al. 2002)
Mounted chamber	1 mL	*E. coli*	Milk sample	8	8	50 mV	1-1 MHz	Cr + Au	Glass	7×10^2	Z	10 Hz	7 h	(Liu et al. 2015)
Dishes	2,5 mL	*E. coli*	LB medium	15	15	100 mV	20-100 kHz	Ta-Ni-Pt	Ceramic	10^{10}	Z	-	12 h	(Zikmund et al. 2010)
Mounted chamber	8 mL	*P. aeruginosa*	TSB	10	10	10 mV	1-100 kHz	Cr + Au	Silicon	10^8	Cdl	100 Hz	1 h	(Kim et al. 2012)
Dishes	300 μL	*S. epidedrmidis*	TSB	20	30	100 mV	10-1 MHz	Cr + Au	Silicon	10^6	R	10 Hz	4 h	(Jacobo Paredes et al. 2014)

Well-plates	300 µL	*S. aureus* *S. epidedrmidis*	TSB	20	30	100 mV	10-1 MHz	Cr + Au	Silicon	10^6	R	10 Hz	6 h	(Paredes et al. 2013b)
Reactor	300 mL	*S. epidedrmidis*	TSB	20	30	100 mV	10-1 MHz	Cr + Au	Silicon	10^6	C	10 Hz	4 h	(Paredes et al. 2012)
Mounted chamber	µL-mL	*S. epidedrmidis*	TSB	20	30	100 mV	1-1 MHz	Cr + Au	Silicon	10^6	C	10 Hz	6 h	(Paredes et al. 2014)
Microfluidic chamber	µl/min	*S. aureus*	TSB	21	21	-	100-1 MHz	Ti + Ni + Au + (Si_3N_4)	Glass	10^7	Z	10 kHz	~1 h	(Estrada-Leypon et al. 2015)
Microfluidic chamber	µL	*E. coli*	DI water	10	20	-	1 kH-1 MHz	Au	SiO_2	10^3	Z	-	~s	(Jiang et al. 2014)
Mounted chamber	80 µL	*B. subtilis* *S. oneidensis*	KCl buffer	-	-	-	10-300 kHz	Au	Glass	-	Z	30 Hz	5 h	(Bonetto et al. 2014)

to establish the relationship between the parameter variation and the number of microbes growing in the media.

Another important characteristic is the **location and position of the sensor** inside the culturing chamber. In other words, how are the sensors exposed to the microbial culture and how does the flow regime affect them? So, the response of the biosensor might be different, depending on what setup is chosen for the experiment. In this way it is possible to target preferentially biomass detection over metabolic activity.

Figure 5 represents the **volumetric units** of the previously described setups in which bacterial biofilms are depicted according to flow dynamics in the volume. Only the modified CDC reactor presents stirring and therefore bacterial cells are exposed to all surfaces for attachment. In the case of multi-well plates and dishes, cells are deposited mostly at the bottom due to gravity. Therefore, the biofilm will be generated mostly at the bottom. This is the same scenario as in the static performance of Microwell on chip where bacteria get deposited at the bottom over the biosensors' surface. And lastly, the microfluidic channels are operated with media flow in most of the cases, so bacteria are exposed to all the inner surfaces of the microchannel. A biofilm will be generated homogeneously along the channel.

Impedance monitoring of biofilm growth in these 5 configurations will be affected by the way in which the sensor is exposed to bacterial cells as discussed earlier. Therefore, there are different electrical parameters that are more sensitive in each scenario. A complete and detailed review is presented in the following sections.

It is important to note the increasing role of **microfluidic devices** during the last years for biofilm research. The unique characteristics and effectiveness of these small channels make them very attractive to conduct experiments focused on different applications, such as evaluation of the effect of surface roughness on bacteria attachment, evaluation of new antimicrobials or antifouling treatments and the increasing resistance of bacteria to these treatments (Pires et al. 2013), screening novel strategies for destabilizing the extracellular matrix or other desegregating methods, combination of electric or magnetic fields as a tool to increase treatment efficiency, etc.

In this context of microfluidic devices, complex geometries or sophisticated detection systems can be implemented by adding some extra components, such as membranes for bacterial capture (Jiang et al. 2014). Definitely there is a great interest in new approaches to develop point of care (PoC) analysis systems for applications in the field of water management, health care or food industry.

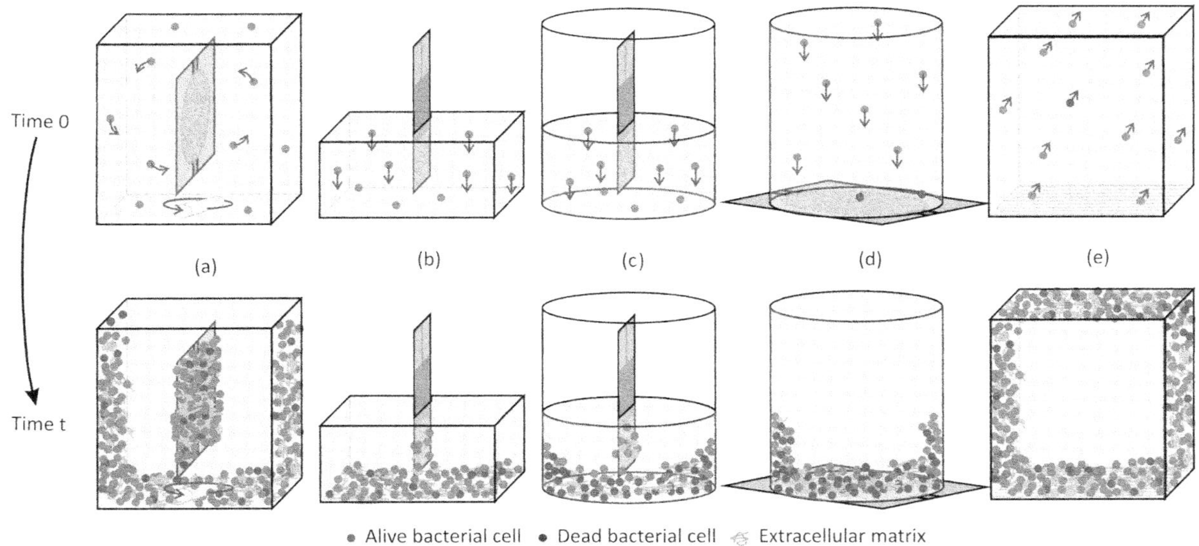

Figure 5. Scheme of a volume unit of each biosensor in the described setups. Correspondence (a) Modified CDC Reactor (b) Petri dishes, (c) Multi-well plates, (d) Microwell-on-chip and (e) Microfluidic device (Paredes et al. 2014).

It is especially remarkable that in such small volumes the signal-to-noise ratio is maximized and therefore the sensitivity of the devices increases. If using reagents, the volume needed is very small and so is the cost of some experiments, etc.

2.3 Electrodes design and configuration

Electrodes or microelectrodes are really working as transducers of the biosensor either decorated for specific bacterial detection of surface-free to monitor changes of conductivity caused by biofilm development. The geometry of the microelectrodes is key to maximizing the sensitivity of the biosensor. One can distinguish basically between macro (wires, plates, etc.) and microelectrodes that are two-dimensional structures over a substrate.

In this section the authors focus on discussing one special type of microelectrodes widely used for biofilm monitoring: Interdigitated microelectrodes (Yang and Bashir 2008). These are common structures to perform impedance measurements. It is a simple 2D geometry with a **capacitive character** that provides high ratio signal/noise, high sensitivity, high stability and large sensitive areas. The main drawback of these 2D structures is that the sensible volume is limited by the design of the microelectrodes, that is, the width and space of the microelectrodes determine the high at which the electric field is confined (Paredes et al. 2012, Varshney and Li 2009).

Figure 6 presents two common geometries for dielectric impedance monitoring (left) and for electrochemical impedance monitoring (right). The difference between them is that electrochemical measurements require to fix the measurement potential and therefore require one electrode for that purpose. For electrochemical measurements as well, the design of the geometry of the electrodes is free, so it is possible to have interdigitated microelectrodes plus an extra electrode to fix the electric potential of the cell.

In the case of interdigitated microelectrodes the size (that is the width and separation of the digits) should be designed according to the size of the target and the measurement method. According to Couniot et al. the microelectrodes width and gap should be in the range of 4 times the size of the microbe target (Couniot et al. 2013). In the case of biofilm formation, it is even more complex since the target changes its size over time. Decisions are usually taken based on experience and experimental results sorting what is the best design to maximize the sensitivity. More simulations of sensitive area before experimentation can highlight interesting design specifications useful for maximizing the sensitivity and the detection limit.

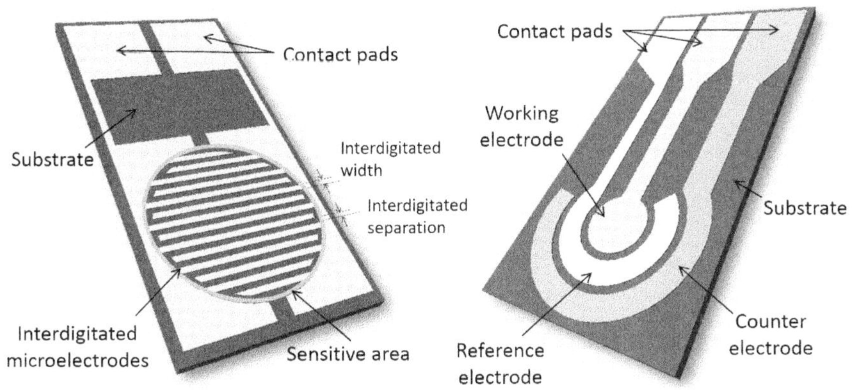

Figure 6. Schematic of an interdigitated microelectrode biosensor (geometry and parameters) and a common electrochemical sensor with three electrodes (working, counter and reference).

Regarding the fabrication materials, all must be biocompatible and non-reactive. Silicon and glass are the main materials for the substrate of the biosensors while gold, platinum and silver are used for the microelectrodes (ITO, Indium Tin Oxide, and others as well).

Equation (3) shows the mathematical expression of a simplified Resistive-Capacitive (RC) serial circuit that could represent the interdigitated microelectrode geometry regarding the resistive and capacitive effects.

$$\vec{Z} = Z_R + Z_C = R + \frac{1}{CWj} \tag{3}$$

where Z_R and Z_C are the impedimetric real and imaginary components of the impedance and R and C are the resistance and capacitance of the equivalent circuit. w is the angular frequency ($w = 2\pi f$) of the sinusoidal signal that is used for exciting the sample. Therefore, the response of this type of sensors will vary along the frequency range.

Figure 7 presents a general response (Bode diagram) of the impedance magnitude and phase of an interdigitated microelectrode sensor in an infected culturing medium. The magnitude decreases when increasing the frequency due to the lower influence of the capacitive term of the impedance. Also it is possible to describe three different regions in the range of the frequency according to the phase of the impedance. There is a capacitively dominated region followed by a combined resistive and capacitive behavior and finally at high frequencies the resistance dominates the impedance response (Paredes et al. 2013b).

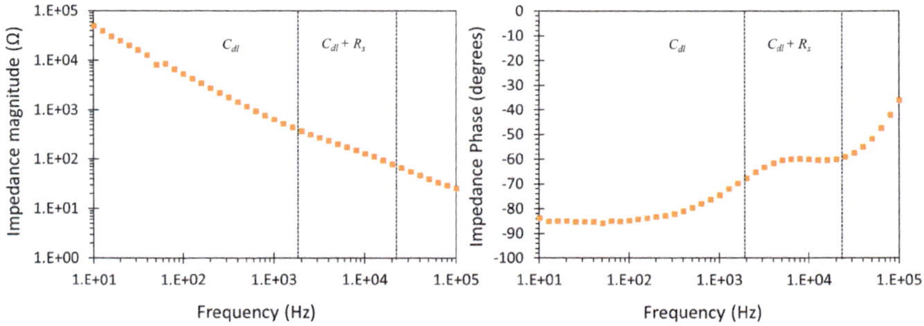

Figure 7. Bode plot of the typical impedimetric response of an interdigitated microelectrode biosensor (magnitude and phase) in culturing media in the domain of the frequency. Typical range 1 Hz to 1 MHz.

2.4 Sensitivity

Sensitivity refers to the number of colony-forming units a biosensor can detect per mL (CFU/ml). Obviously this parameter will depend on the selected method for detection as well as the sample homogeneity and the detection principle. Special care is required when comparing sensitivities of different publications in order to understand what detection principle has been applied.

Figure 8a represents the usual **growth kinetics curve of a microbial culture** in four differentiated phases: lag, exponential, stationary and death. The lag phase is critical for the detection time as the impedance changes requiring a certain metabolic activity (and subsequently an increase of the number of bacterial cells). This trend is also applicable to biofilm growth although these non-flocculant bacteria grow attached to free surfaces. Impedance variations will be proportional to the number of microbes present in the sample. Therefore, it is possible to determine the maximum sensitivity of a biosensor by comparing relative changes with the increase of microbes. In this case the authors are leaving out the distinction between effects provoked by metabolic activity or biomass increase and consider everything as a whole.

In Fig. 8 is represented the detection time calculated based on the changes in the trend of the relative changes of the impedance. Researchers use either a threshold or the change of the slope to determine whether the sample is infected or not. After the first detection, the trend is still analyzed to confirm the positive changes due to bacterial presence.

Figure 8b represents the detection time vs. the initial concentration of bacterial inoculum in the media. The graph includes many of the papers that are reviewed in this chapter. Lines represent the detection limit tested by

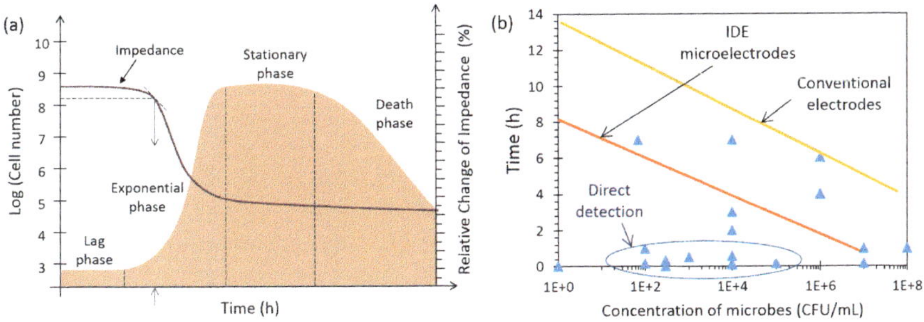

Figure 8. (a) Representation of the growth kinetics of a microbial culture divided in its different phases. Impedance variation is represented according to the increase of biomass. (b) Representation of the detection time vs. the initial concentration of microorganisms.

Yang and Bashir (2008) in their paper. They were comparing conventional electrodes and interdigitated microelectrodes achieving a higher sensitivity in the second case. Also the detection limit was first calculated by Gomez et al. for metabolic activity detection inside a microchamber (Gomez et al. 2001).

There are two groups of points in the figure corresponding to the two described detection methods: direct detection of bacterial cells (marked in the figure) and biofilm monitoring. The detection time of this first group is less than one hour as it is rapid detection; the second group corresponds to a scatter map with detection times ranging in a few hours.

2.5 Detection strategies

There are a number of methods for detecting bacterial presence either directly targeting the **presence of bacterial cells or biofilm development** over surfaces. The first approach usually requires to take a sample off and transfer it to the analysis device. However, it is also possible to use these devices *in situ* with certain limitations like lifespan, stability, reuse, etc. Certainly detecting biofilms starts by detecting bacterial cells before and while biofilm development.

Biofilm monitoring usually requires measuring *in situ*: pipelines, deposits, catheters and any exposed surface in general, for prolonged periods of time. Biosensors are exposed to the sample for monitoring and therefore are exposed to the exact same conditions, temperature, cleaning processes, changes in media, etc. Biosensors therefore are required also to withstand very demanding conditions in many occasions.

2.5.1 Bacterial cells detection

Specific detection of bacteria requires a biorecognition element, usually antibodies (but also DNA, proteins, etc.), which can be used in different ways: immobilized over the microelectrodes (Radke and Alocilja 2005, Tang et al. 2011), conjugated with magnetic micro beads (Varshney and Li 2007, Yang 2008). Also there are other strategies, for example using the dielectrophoresis effect (DEP) for cellular concentration and trapping and following detection (Gómez-sjöberg et al. 2005, Suehiro et al. 2006, 2003).

These approximations target the detection of the presence of bacteria in a specific time whether or not they are generating a biofilm. The ultimate goal is to provide high sensitivity (due to the small concentration of microbes), high selectivity and fast response. These types of biosensors are ideal for applications in which it is possible to have the sample to be analyzed. However, it is not possible to integrate them in long-term applications.

Figure 9 represents the main cellular detection methods for direct detection of bacteria: surface-free, dielectrophoresis effect, Self-Assembly Monolayer (SAM) of antibodies and decorated magnetic nanoparticles.

The first detection method is based on the disturbance of the electric field caused by the presence of 'particles' on top of microelectrodes. This option has very little sensitivity and even worse selectivity. The use of the dielectrophoresis effect drags bacteria in an undifferentiated manner to the surface of the biosensor. It is acting as a collector for performing the impedimetric measurement. The result of this method is highly affected by the charge characteristics of the bacteria and the media conductivity. Also, it is extremely important to control the electrical parameters, such as voltage, current and frequency and avoid effects like hydrolysis that would interfere with the measurement. Suehiro et al. (2003) suggest that low frequency electric fields achieve better results than higher frequencies. These first two methods are cheap and easy to integrate *in situ* in more complex systems.

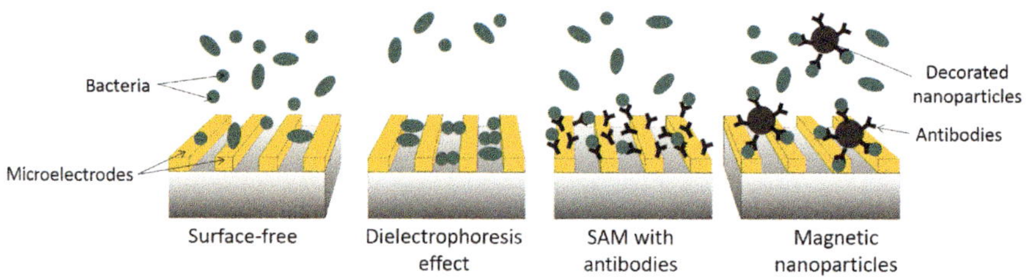

Figure 9. Schematic representation of the biosensors performance using different methods for direct detection of bacterial cells.

The following two methods are highly specific and sensitive. Antibodies can be immobilized on the microelectrode surfaces using a molecular stack (self-assembled monolayer). These type of sensors present high selectivity and sensitivity, but are expensive and difficult to optimize. Some techniques use a secondary molecular signal to amplify the sample detection, such as lectins (these are small, carbohydrate-binding proteins that can reversible attach to the cellular membrane due to the high specificity for sugars) that get attached to the surface of the bacterial cells increasing the response measured (Li et al. 2015).

A variation of these concepts is the last depicted method in which the antibodies are immobilized on the nano-particle surface. After the immunoreaction with the sample, these nanoparticles are concentrated, using magnetic fields on the sensors' surface and increasing the signal of the measurement.

2.5.2 Biofilm detection and monitoring

There are mainly three different approaches for bacterial biofilm detection and continuous monitoring for *in situ* applications: Measuring changes in conductivity due to biomass generation, metabolic activity and substrate uptake or consumption. Figure 10 represents a schematic and simplified way of these three scenarios in biofilm monitoring.

In the first case, impedance changes are caused by an increase of the biofilm volume, bacterial cells and the extracellular matrix. This method is appropriate for situations in which the liquid is flowing and the medium presents steady parameters (no changes in conductivity) due to renewal. The second method is based on monitoring conductivity changes in the medium

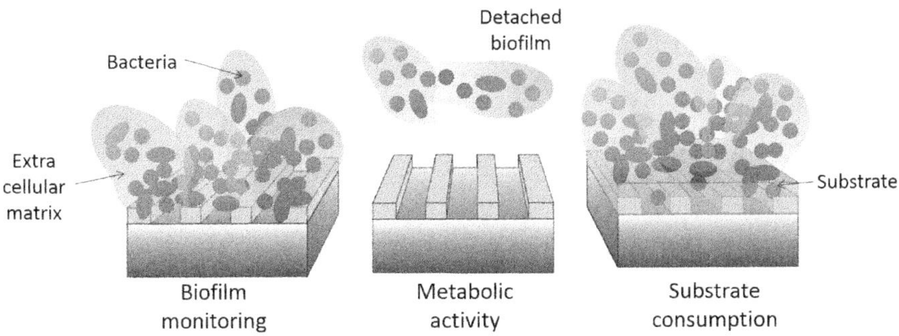

Figure 10. Schematic representation of the biosensors performance using different methods for biofilm monitoring.

caused by metabolic activity of biofilms, regardless of the attached bacteria. In scenarios where there is no renewal of the medium, these changes will be more significant than in the previous scenarios.

It is obvious that it is not possible to uncouple these to situations of biomass increase and metabolic activity. Also it is important to mention that real applications do not present homogeneous media and therefore the location of the biosensor will balance the effect of biomass coating over changes of media or vice versa. This point will be discussed in some more detail in the following sections.

Usually these two approximations are coupled and therefore is not possible to discard one over the other. In fact, Pires et al. (2013) demonstrated that it is possible to combine different techniques to evaluate both biomass development and metabolic activity in the same sample. Furthermore, these authors used sodium azide (which is a compound commonly used for biofilm treatment) as a treatment over a mature biofilm. The result shows that while current values (and later trend) decreased, impedance values remained fairly stable. This suggests that bacteria were killed but the extracellular matrix and the rest of biomass still adhered to the sensor.

The third option is based on substrate consumption of specific nutrients placed on the biosensors microelectrodes. This substrate acts as a passive layer that will be modified or consumed by the metabolic activity of the bacteria attached. It happens to be an interesting approach based on the affinity of certain groups of bacteria to specific molecules that only they can metabolize.

2.6 Comparison of detection strategies

Selection of biosensor, detection strategy and measurement performance will depend on the specific application. In this section the authors present a comparison between the previously described methods. Table 2 shows a compilation of some of the recent advances in the field for different applications. The table is divided into 4 different groups regarding the detection principle: antibodies (and other immobilized molecules), conjugated magnetic particles, dielectrophoresis effect and surface-free transducers.

Electric parameters represented in the table are: Impedance magnitude (Z), electrochemical impedance (*EIS*), electrical conductivity (ρ) capacitance (C) and an equivalent electrical parameter double layer capacitance (C_{dl}).

Table 2. Comparison of different setups for biofilm cultivation.

Device	Max. volume	#biosensor/assay	Agitation/stirring	Media flow/ dynamic	Max. assay duration[a]	Direct visualization
Reactor	~350 mL	18/Reactor	Allowed by stirring	Allowed	No limit	Not allowed
Dishes	~50 mL	8/Dish	Allowed	Allowed	No limit	Not allowed
Well-plates	200–300 µL	8x4/Plate	Allowed	Not allowed	< 50 hours	Not allowed
Chamber	Few mL	1/Device	Not allowed	Allowed	No limit	Allowed
Microfluidic device	Several µL	#/Device	Not allowed	Allowed	No limit	Allowed

[a] Approximate values according to experimental results.

Table data confirm what has been presented earlier about the two main approaches for biofilm detection and monitoring—either direct bacterial detection or monitor changes caused by biofilm growth. Most of the applications were developed for food and water applications, in which rapid detection and selectivity are key.

On the one hand, the first three groups use biorecognition elements aimed at a fast and specific cellular detection before bacteria develop biofilms. These strategies seek to increase the number of binding events with the sample by maximizing the exposure of the sample to the sensor using magnetic concentration or dielectrophoresis effect. Selective detection is achieved by targeting specific binding of one type of microorganisms (using the right antibodies). The detection limit of these approaches is in the range of 10^0 to 10^5 CFU/ml, depending on the technology and the sample. Also, detection times are in the order of seconds to a few minutes, which is the required time for the sample to bind to the recognition elements. There are also a couple of papers that describe the ability to detect bacterial cells using surface-free microelectrodes inside a micro-chamber with a detection limit around 10^3–10^4 CFU/ml (Jiang et al. 2014, Yang 2008).

On the other hand, the references in the table using surface-free microelectrodes show that this approach is appropriate for biofilm development monitoring. Changes in impedance (or equivalent electrical parameters) serve to monitor how biofilms are being formed over the surface of the biosensors. Initial concentrations collected from these papers show values around 10^5 to 10^8 CFU/ml. However all the authors in this section agree that only one bacterium is necessary to develop a biofilm and therefore its formation is just a matter of time. The key point is that the biosensor in conjunction with the biofilm kinetics provides enough change to detect it in an early stage of development.

Another interesting result of the analysis is that surface-free monitoring presents its maximum variation at a low frequencies. The following sections review how the signals are affected by the exciting frequency. On the contrary when using antibodies, a middle range of the frequency provides the highest sensitivity. In a similar way, the DEP (Dielectrophoresis) effect uses a high frequency range to induce cellular movement over the electrodes under an electric field.

2.7 Equivalent circuit

Equivalent electric circuit are models used for the determination of specific electric parameters of each biological effect on the total measured impedance. For example, the medium conductivity characteristics and its variations due to metabolic activity, the increase of biofilm attached to the surface, the aggregation of bacterial cells and other effects, such as ion movement or redox effects. Modeling electrical parameters only makes sense for monitoring biofilm development. It is important to know what kind of effect is dominating over the others for directly analyzing the biomass coating on the surfaces. On the contrary direct detection of bacteria usually does not apply these models to cellular identification.

Latest research tends to include these types of analyses not only to better understand the biological behavior but also to maximize the biosensor performance. Each model can be expressed as an equation of the behavior of the system in the frequency domain and used for fitting experimental data to obtain the values of each equivalent electrical component. In this sense it is possible to analyze independently the trend of each biological or chemical effect listed before.

Common parameters introduced in the electrical model are: dielectric resistance and capacitance of the medium (R_{sol} and C_{di}, respectively), a double layer capacitance (C_{dl}) that expresses the ionic charge over the electrodes, the capacitance and resistance caused by the biomass over the sensor (R_b and C_b). There are also models that include the capacitance of the bacterial membrane (C_m) or the resistance of the cytoplasm (C_c). In the case of electrochemical sensing, there are additional parameters usually included that express the electronic transfer between the sample and the microelectrodes: Constant phase elements (*CPE*) and more specifically, the Warburg impedance (Z_w) and the electron-transfer resistor (R_t).

There is still not full agreement on which type of electrical model is best to represent the complex biological problem. One of the reasons for this lack of understanding is that every system used in research is unique and that is very difficult to agree on the degree of complexity that every author wants to give to the system. Figure 11 represents some of the circuits found in the literature regarding biofilm monitoring. A brief description is included for each circuit highlighting the most relevant features.

Equivalent circuits are sorted by increasing complexity of the system or the number of elements that are included. Fitting analysis is performed using experimental data in the frequency domain. It is important to analyze not only the values of each parameter but also how the frequency influences the behavior of the measurements. By repeating this analysis at different times it is possible to analyze the changes of the specific electrical parameters of the model during the biofilm growth.

(A)

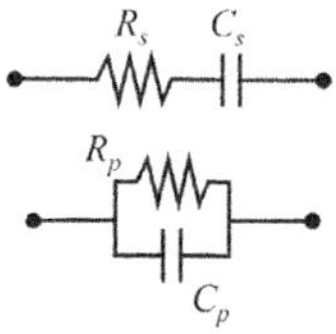

Maximum simplification of the electrical circuit by extracting the serial (or parallel) equivalent parameters of the resistance and capacitance regardless the biological effect. Could be used to increase the sensitivity in a very easy way with the ultimate goal of being integrated in an electronic control circuit (Paredes et al. 2012).

(B)

A variation of the previous circuit but taking into account that the majority of the capacitance effect (specially at low frequencies) is caused by the C_{dl}. (Liu et al. 2015, Paredes et al. 2014, Varshney and Li 2008, Yang 2008).

(C)

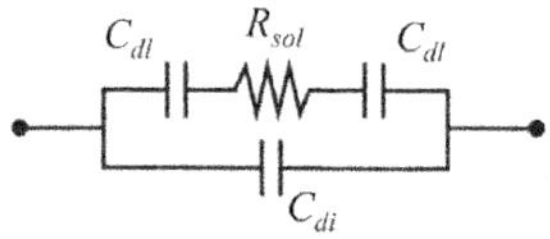

This model focusses on the media (R_{sol} and C_{di}) and the double layer capacitance (C_{dl}). Yang et al. (Yang et al. 2004) are measuring mainly the metabolic activity in a sample with planktonic bacteria. But also a number of authors used this the same approach too (Radke and Alocilja 2005, Varshney and Li 2007). Probably this is the most used equivalent circuit in bibliography.

(D)

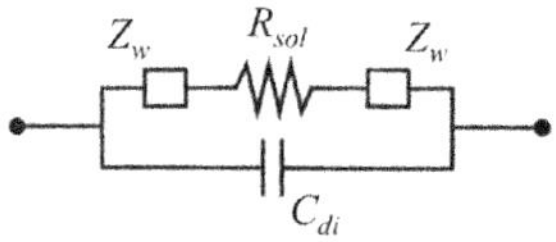

A variation of the previous circuit replacing the capacitors by constant phase elements such as a Warburg element (Gomez et al. 2002, 2001, Varshney and Li 2007). At the lower frequencies Z_w dominates the response showing the capacitive behavior of the biosensor.

(E)

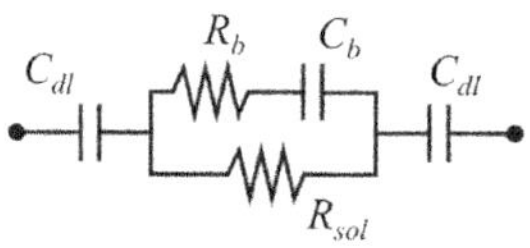

The double layer capacitance is one of the most relevant elements for biofilm monitoring. Also in the circuit there are two branches accounting for the bulk resistance of the media and the biofilm components (by a resistance and capacitance).

(F)

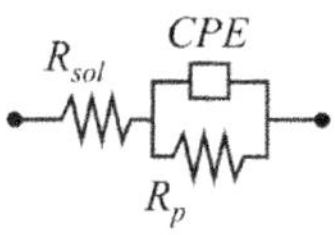

This circuit consists of a serial resistance with a charge transfer resistance (or polarization) in parallel with a constant phase element that describes the effect of a real capacitance (phase 45°) (Maalouf et al. 2007, Oliver et al. 2006).

(G)

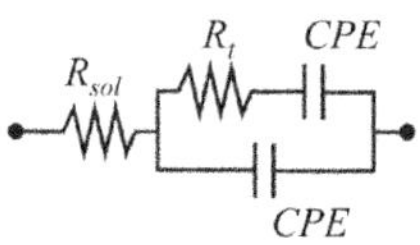

Resistance of the solution in series with a couple of constant phase elements (regarding the capacitance of the double layer and the biomaterial absorption) and an electron transfer resistance. This circuit model is mostly used when having electrochemical reactions (Jiang et al. 2014, Kim et al. 2012, Li et al. 2015, Yang and Li 2005).

(H)

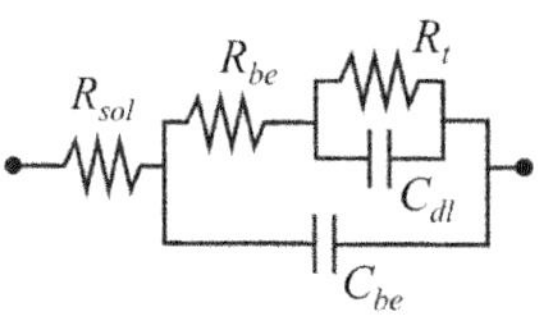

Bayoudh et al. propose a couple of equivalent circuits, this is the more complex one focused on the interfaces that exchange charge (bacterial electrode and electrode electrolyte). The parameters C_{dl} and R_t are affected by the cellular adhesion over the electrodes showing how bacteria changes the interface (Bayoudh et al. 2008).

(I)

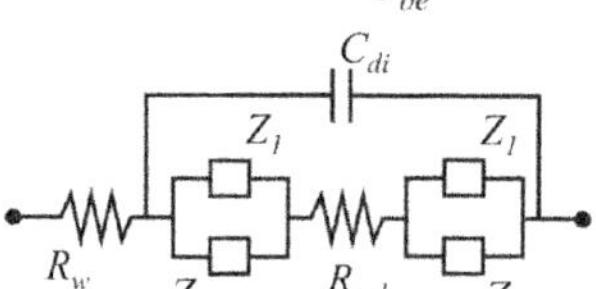

Like in the previous model, Gómez-sjöberg et al. presented an equivalent circuit focused on the electrical or electronic interfaces (Z_1 and Z_2). Then the resistance of the chip (R_w) dielectric properties of the environment (C_{di}) and finally the bulk resistance of the media (R_{sol}) (Gómez-sjöberg et al. 2005). Radke et al. use a slightly modification but with the same principle (Radke et al. 2004).

Figure 11. Equivalent circuits used for fitting experimental data in biofilm monitoring research.

2.8 Monitoring bacterial biofilm development

Monitoring usually is based on the relative changes of the magnitude during the microbial culture and biofilm development. Different ranges of the frequency band are more affected by some biological processes during the measurement. There is a broad agreement about the use of the low range of frequency for biofilm monitoring. It is precisely at this range (1 Hz–500 Hz) at which the double layer capacitance dominates the impedimetric response (Varshney et al. 2007, Yang 2008). At higher frequencies, on the contrary, the resistance is the parameter that dominates the response (500 Hz–100 kHz). Many authors agree with this analysis and propose the use of the low range of the frequency for biofilm monitoring, regardless of the detection strategy selected for this purpose (Paredes et al. 2014, Radke and Alocilja 2005, Tang et al. 2011, H. Wang et al. 2012).

The double layer capacitance is one of the more sensitive parameters for biofilm detection as it evaluates the interface electrode biofilm during the experiment. Many authors use this parameter in the equivalent model to evaluate the increase of biomass attached to the biosensors' surface. Experimental data is shown in the following graphs to explain in more detail these statements.

Figure 12 shows the relative changes of the impedance magnitude in four different setups described in the previous sections: CDC Reactor, petri dishes and well-plate and on-chip chamber. Experimental conditions are the same in the four setups so that it is possible to compare them with each other. It is clear that the lower range of the frequency the bigger the relative variations on the impedance magnitude for all selected setups. However, there are two that show very little variations compared to the other two and that could be explained by the position of the sensor in the culturing media. Bigger variations in the relative changes are associated with a higher coverage of the sensor with biofilm while for smaller changes, the coverage is not so broad.

Figure 13 shows precisely the relative variations of different equivalent parameters in the same setups—equivalent resistance and capacitance instead of the impedance magnitude. These are not equivalent circuit components but another way to express impedance characteristics. The differences between the discussed setups are even more obvious. Vertical sensors that are not fully in contact with biofilms show higher variations in the resistance regarding the metabolic activity, which is reflected in changes in conductivity. On the contrary, the sensors that are disposed so that a biofilm is generated over their surface show higher variations in capacitance (related to the double layer capacitance) and apparently no changes in the resistance.

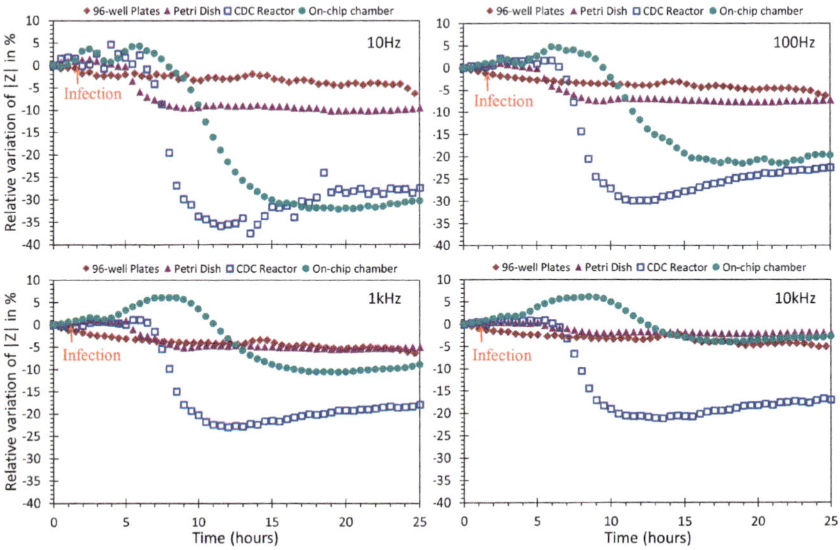

Figure 12. Relative variation of impedance magnitude of a culture of *Staphylococcus epidermidis* (ATCC 35984 strain) in culture media at 4 different frequencies, with an initial inoculum of ~1 x 10^6 CFU/ml. The impedance response of four different setups is represented in the graphs (Paredes et al. 2014).

So it is clear that there is a relationship between the selected setup for monitoring bacterial biofilm development and the electrical parameters that are more affected by either biomass increase or metabolic activity. Increase of the attached biofilm on the biosensor will cause changes in capacitance and impedance magnitude, while metabolic activity will change the conductivity of the media and therefore will have a reflection on the resistance.

According to the equivalent electric models presented before these two behaviors could be as well translated to specific elements in those models increasing even more the sensitivity of the system. Many of the reviewed works focused on the capacitive parameters as the ultimate goal is monitor biofilm attachment and growth.

3. Examples of Specific Applications

This section briefly focusses on two specific fields and applications of the technology for bacterial or biofilm detection. The first one concentrates on the healthcare environment, more specifically on central venous catheters; the second on the food industry environment, analyzing the detection limits

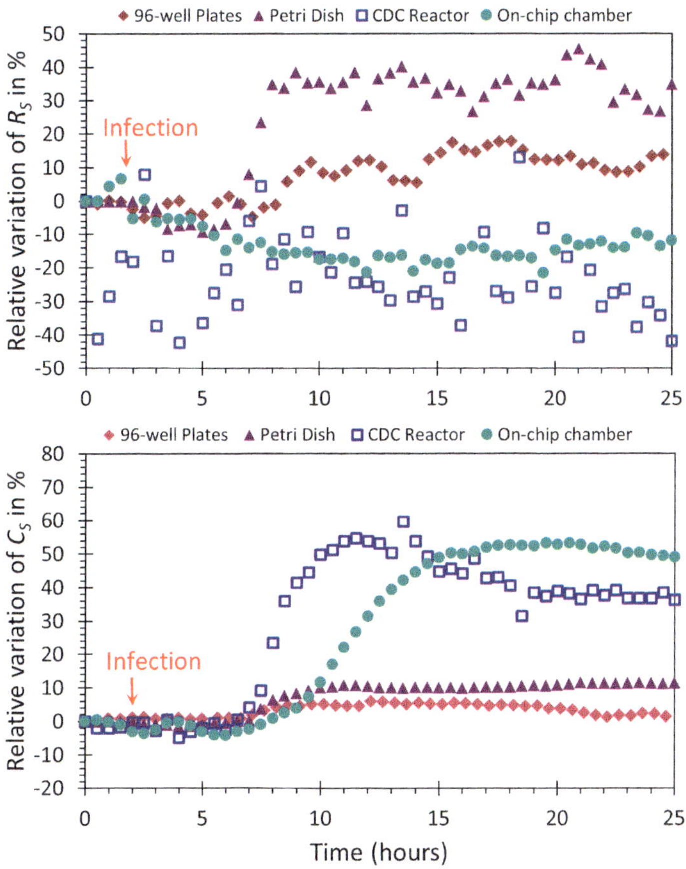

Figure 13. Relative variation curves of equivalent serial resistance (a) and equivalent serial capacitance (b) during a *Staphylococcus epidermidis* (strain ATCC 35984) biofilm culture in the four culturing setups. Both panels measured at 10 Hz with initial inoculum of ~1 x 10^6 CFU/ml (Paredes et al. 2014).

and response time, technique limitations and other important issues of the current technology. However, there are many more promising applications that are currently under development that hopefully will become a reality very soon.

3.1 Healthcare environment

Central venous port catheters are just an example of a number of medical devices that present medical complications due to biofilm colonization. They all share the intrinsic risk of infection during the medical procedures.

Eighty per cent of human infections are related to biofilm formation and 50 per cent of those are acquired in the hospital environment. Diagnosis of these cases are usually too late for early treatment, causing a great cost economically and for the patients' health.

There are a number of developments that suggest the integration of biosensors based on microelectrodes to detect the formation of biofilms at early stages, so that the medical outcome would improve dramatically (Ehrlich et al. 2006). Wilson and Moragan made another proposal for biofilm detection over the surface of hip replacements (Wilson and Moragan 2010). Another interesting field is the cardiovascular one as many devices are largely exposed to these complications (Baddour et al. 2012). Unfortunately there is still a long way to go to make available this technology within the medical devices.

Paredes et al. presented in 2014 a new central venous port integrating a biosensor and every other electronic module required to control the monitoring system (Paredes et al. 2014). Such devices are subjected to biofilm infection during their normal operation by qualified medical personnel. Infection rates vary from different studies, but some estimations point from 0.1 to 0.86 episodes per every 1000 days of use, or around 3 per cent of the implanted ports (Maki et al. 2006, Ng et al. 2007).

Figure 14 presents a CAD image of a prototype of a Central Venous Port integrating all the modules from the biosensor to communications,

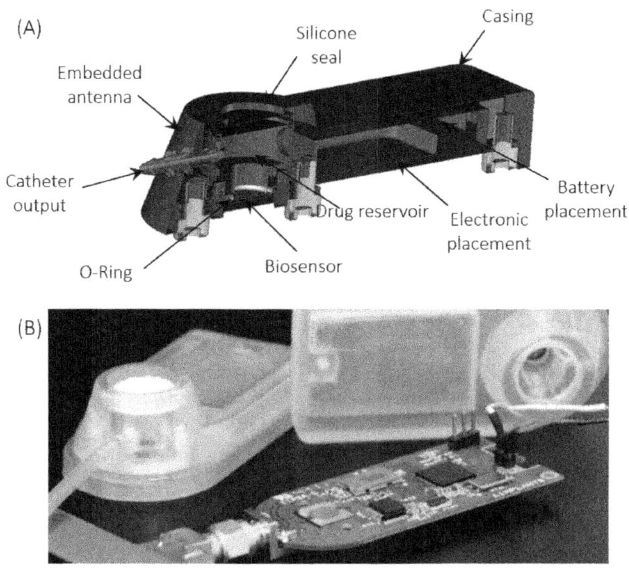

Figure 14. (A) CAD image of the prototype of a central venous port with all the modules and (B) a real picture of the fabricated prototypes, including electronic board and biosensor.

electronics and battery. Also there is a real picture of the developed prototype. Different tests were performed over the device, from communication performance (achieving around 6 m of distance communications from a phantom of the body chest) to microbiological cultures. The device passed successfully all the experiments but still needs to increase selectivity and sensitivity in order to ensure the high quality standards required in the healthcare system (Paredes et al. 2013a).

3.2 Food industry

Pathogen detection and identification in the food industry is still a challenge with a great impact on public health and the economy (Yang and Bashir 2008). According to the authors, diseases caused by microorganisms found in food cost over 76 million dollars/year, more than three hundred hospitalizations and about five thousand deaths every year in the United States alone. Among all the microbes involved in food contamination, *Salmonella enteritidis* and *Escherichia coli* are probably the most feared pathogens and definitely the most studied ones (Kim et al. 2008, Pedrero et al. 2009, Yang et al. 2003). *Escherichia coli* is also an indicator of fecal contamination.

Along this chapter, numerous works have been cited in this field presenting solutions for this challenge. Current methods for analysis are slow and costly and the access to samples is not always easy, so they do not reach the sector expectations. In this context and over the last few decades, new detection methods have been published to achieve the demanding requirements. These new procedures and techniques are faster (in the order of hours) and very precise (detecting 10^3 to 10^4 CFU/ml). However, there is still a need for improvement in both sensitivity and selectivity and overall, a better integration of these biosensors directly in the production lines.

The reality is that even though there is a great need for this technology and great advances in techniques and devices has occurred, there is not much transfer to commercial devices available for the companies. Industry still relies on lab work for the analysis of samples from their production (H. Wang et al. 2012).

4. Conclusions

Impedance spectroscopy has been largely used for bacterial detection and biofilm monitoring and its efficacy has been proven in many different devices and applications. However, there is still a lack of translation from academic research to real applications and devices.

This chapter presents a complete review of the different setups that researchers have developed for biofilm research. Each one mimics different real conditions and provides very interesting results about specific biological behaviors. Location and position of biosensors are key to understand the monitoring outcome. Therefore, real *in situ* integration of biosensors should be analyzed carefully to obtain the best possible performance of the biosensors. Also, electrical equivalent circuits have been applied to gather more knowledge about specific behavior of the biological system. Details of the most common circuits have been discussed in this chapter as well.

Different detection strategies have been presented for early direct detection of bacterial cells and for real-time monitoring of biofilm development. Direct detection techniques show high performance for rapid and selective detection of microbes. These techniques require sample extraction and in many cases some sample preparations. Microfluidic devices are a promising platform that allows the research of many features related to biofilm performance. In the coming years, we will see more research projects using these technologies and proposing novel approaches. Finally surface-free biosensors have been proposed for long-term applications, showing that it is possible to detect biofilm development from early stages.

In summary, impedance monitoring has proven a suitable technique for biofilm detection and monitoring and we can be sure that more biosensors will be applied in real-field occasions.

5. Acknowledgements

The authors of this chapter want to acknowledge the editors and reviewers' effort in editing this book and are thankful for the invitation to participate with this chapter.

Keywords: Impedance microbiology, Interdigitated microelectrode biosensor, Bacterial biofilm, *in situ* bacterial detection, Label-free biosensor

References

Baddour, L.M., Cha, Y.-M. and Wilson, W.R. 2012. Clinical practice. Infections of cardiovascular implantable electronic devices. N. Engl. J. Med. 367: 842–9. doi:10.1056/NEJMcp1107675.

Blenkinsopp, S.A. and Costerton, J.W. 1991. Understanding bacterial biofilms. TIBTEC 9.

Cady, P., Dufour, S.W., Lawless, P., Nunke, B. and Kraeger, S.J. 1978a. Impedimetric screening for bacteriuria. J. Clin. Microbiol. 7: 273–8.

Cady, P., Dufour, S.W., Shaw, J. and Kraeger, S.J. 1978b. Electrical impedance measurements: rapid method for detecting and monitoring microorganisms. J. Clin. Microbiol. 7: 265–72.

Coenye, T. and Nelis, H.J. 2010. *In vitro* and *in vivo* model systems to study microbial biofilm formation. J. Microbiol. Methods 83: 89–105. doi:10.1016/j.mimet.2010.08.018.

Costerton, J.W., Stewart, P.S. and Greenberg, E.P. 1999. Bacterial biofilms: a common cause of persistent infections. Science 284: 1318–22.

Couniot, N., Flandre, D., Francis, L.A. and Afzalian, A. 2013. Signal-to-noise ratio optimization for detecting bacteria with interdigitated microelectrodes. Sensors Actuators, B Chem. 189: 43–51. doi:10.1016/j.snb.2012.12.008.

Davies, D. 2003. Understanding biofilm resistance to antibacterial agents. Nat. Rev. Drug Discov. 2: 114–22. doi:10.1038/nrd1008.

del Pozo, J.L., Rouse, M.S., Mandrekar, J.N., Sampedro, M.F., Steckelberg, J.M. and Patel, R. 2009. Effect of electrical current on the activities of antimicrobial agents against *Pseudomonas aeruginosa*, *Staphylococcus aureus*, and *Staphylococcus epidermidis* biofilms. Antimicrob. Agents Chemother. 53: 35–40. doi:10.1128/AAC.00237-08.

Dheilly, A., Linossier, I., Darchen, A., Hadjiev, D., Corbel, C. and Alonso, V. 2008. Monitoring of microbial adhesion and biofilm growth using electrochemical impedancemetry. Appl. Microbiol. Biotechnol. 79: 157–64. doi:10.1007/s00253-008-1404-7.

Donlan, R.M. 2002. Biofilms: microbial life on surfaces. Emerg. Infect. Dis. 8: 881–90. doi:10.3201/eid0809.020063.

Donlan, R.M. and Costerton, J.W. 2002. Biofilms: survival mechanisms of clinically relevant microorganisms. Clin. Microbiol. Rev. 15: 167–193. doi:10.1128/CMR.15.2.167.

Ehrlich, G.D., Stoodley, P., Kathju, S., Zhao, Y., Mcleod, B.R., Balaban, N., Hu, F.Z., Sorereanos, N.G., Costerton, J.W., Stewart, P.S., Post, J.C. and Lin, Q. 2006. Engineering Approaches for the Detection and Control of orthopaedic biofilm Infections. Clin. Orthop. Relat. Res. 02418: 59–66.

Estrada-Leypon, O., Moya, A., Guimera, A., Gabriel, G., Agut, M., Sanchez, B. and Borros, S. 2015. Simultaneous monitoring of *Staphylococcus aureus* growth in a multi-parametric microfluidic platform using microscopy and impedance spectroscopy. Bioelectrochemistry 105: 56–64. doi:10.1016/j.bioelechem.2015.05.006.

Giladi, M., Porat, Y., Blatt, A., Wasserman, Y., Kirson, E.D., Dekel, E. and Palti, Y. 2008. Microbial growth inhibition by alternating electric fields. Antimicrob. Agents Chemother 52: 3517–22. doi:10.1128/AAC.00673-08.

Gomez, R., Bashir, R., Sarikaya, A., Ladisch, M.R., Sturgis, J., Robinson, J.P., Geng, T., Bhunia, A.K., Apple, H.L. and Wereley, S. 2001. Microfluidic biochip for impedance spectroscopy of biological species. Biomed. Microdevices 201–209.

Gomez, R., Bashir, R. and Bhunia, A.K. 2002. Microscale electronic detection of bacterial metabolism. Sensors Actuators A Phys. 86: 198–208.

Gómez-sjöberg, R., Morisette, D.T., Bashir, R. and Member, S. 2005. Impedance microbiology-on-a-chip: microfluidic bioprocessor for rapid detection of bacterial metabolism. J. Microelectromechanical Syst. 14: 829–838.

Hall-Stoodley, L., Costerton, J.W. and Stoodley, P. 2004. Bacterial biofilms: from the Natural environment to infectious diseases. Nat. Rev. Microbiol. 2: 95–108. doi:10.1038/nrmicro821.

Harris, L.G. and Richards, R.G. 2006. Staphylococci and implant surfaces: a review. Injury 37 Suppl. 2: S3–14. doi:10.1016/j.injury.2006.04.003.

Ivnitski, D., Abdel-Hamid, I., Atanasov, P., Wilkins, E. and Stricker, S. 2000. Application of electrochemical biosensors for detection of food pathogenic bacteria. Electroanalysis 12: 317–325. doi:10.1002/(SICI)1521-4109(20000301)12:5<317::AID-ELAN317>3.0.CO;2-A.

Jiang, J., Wang, X., Chao, R., Ren, Y., Hu, C., Xu, Z. and Liu, G.L. 2014. Smartphone based portable bacteria pre-concentrating microfluidic sensor and impedance sensing system. Sensors Actuators, B Chem. 193: 653–659. doi:10.1016/j.snb.2013.11.103.

Kim, G., Morgan, M., Hahm, B.K., Bhunia, A., Mun, J.H. and Om, A.S. 2008. Interdigitated microelectrode based impedance biosensor for detection of salmonella enteritidis in food samples. J. Phys. Conf. Ser. 100: 052044. doi:10.1088/1742-6596/100/5/052044.

Knetsch, M.L.W. and Koole, L.H. 2011. New Strategies in the development of antimicrobial coatings: the example of increasing usage of silver and silver nanoparticles. Polymers (Basel). 3: 340–366. doi:10.3390/polym3010340.

Li, Z., Fu, Y., Fang, W. and Li, Y. 2015. Electrochemical impedance immunosensor based on self-assembled monolayers for rapid detection of *Escherichia coli* O157:H7 with signal amplification using lectin. Sensors 15: 19212–19224. doi:10.3390/s150819212.

Maki, D.G., Kluger, D.M. and Crnich, C.J. 2006. The risk of bloodstream infection in adults with different intravascular devices: a systematic review of 200 published prospective studies. Mayo Clin. Proc. 81: 1159–71. doi:10.4065/81.9.1159.

McLandsborough, L., Rodriguez, A., Pérez-Conesa, D. and Weiss, J. 2006. Biofilms: at the interface between biophysics and microbiology. Food Biophys. 1: 94–114. doi:10.1007/s11483-005-9004-x.

Muñoz-Berbel, X., Muñoz, F.J., Vigués, N. and Mas, J. 2006. On-chip impedance measurements to monitor biofilm formation in the drinking water distribution network. Sensors Actuators B Chem. 118: 129–134. doi:10.1016/j.snb.2006.04.070.

Muñoz-Bonilla, A. and Fernández-García, M. 2012. Polymeric materials with antimicrobial activity. Prog. Polym. Sci. 37: 281–339. doi:10.1016/j.progpolymsci.2011.08.005.

Ng, F., Mastoroudes, H., Paul, E., Davies, N., Tibballs, J., Hochhauser, D., Mayer, A., Begent, R. and Meyer, T. 2007. A comparison of hickman line- and port-a-cath-associated complications in patients with solid tumours undergoing chemotherapy. Clin. Oncol. (R. Coll. Radiol). 19: 551–6. doi:10.1016/j.clon.2007.04.003.

Oliver, L.M., Dunlop, P.S.M., Byrne, J.A., Blair, I.S., Boyle, M., Mcguigan, K.G. and Member, E.T.M.S. 2006. An impedimetric sensor for monitoring the growth of, in: proceedings of the 28th IEEE EMBS annual international conference new York City, USA, Aug 30-Sept 3, 2006. pp. 2006–2009.

Paredes, J., Becerro, S., Arizti, F., Aguinaga, A., Del Pozo, J.L. and Arana, S. 2012. Real time monitoring of the impedance characteristics of *Staphylococcal* bacterial biofilm cultures with a modified CDC reactor system. Biosens. Bioelectron. 38: 226–32. doi:10.1016/j.bios.2012.05.027.

Paredes, J., Arana, S., Arizti, F.J., Schmidt, C. and Valderas, D. 2013a. Intelligent Subcutaneous Venous Access Port and Method for Detecting Biolayer.

Paredes, J., Becerro, S., Arizti, F., Aguinaga, A., Pozo, J.L. Del and Arana, S. 2013b. Interdigitated microelectrode biosensor for bacterial biofilm growth monitoring by impedance spectroscopy technique in 96-well microtiter plates. Sensors Actuators B. Chem. 178: 663–670. doi:10.1016/j.snb.2013.01.027.

Paredes, J., Becerro, S. and Arana, S. 2014. Comparison of real time impedance monitoring of bacterial biofilm cultures in different experimental setups mimicking real field environments. Sensors Actuators B Chem. 195: 667–676. doi:10.1016/j.snb.2014.01.098.

Pedrero, M., Campuzano, S. and Pingarrón, J.M. 2009. Electroanalytical sensors and devices for multiplexed detection of foodborne pathogen microorganisms. Sensors (Basel). 9: 5503–20. doi:10.3390/s90705503.

Pires, L., Sachsenheimer, K., Kleintschek, T., Waldbaur, A., Schwartz, T. and Rapp, B.E. 2013. Online monitoring of biofilm growth and activity using a combined multi-channel impedimetric and amperometric sensor. Biosens. Bioelectron. 47: 157–163. doi:10.1016/j.bios.2013.03.015.

Radke, S.M. and Alocilja, E.C. 2005. A high density microelectrode array biosensor for detection of *E. coli* O157:H7. Biosens. Bioelectron. 20: 1662–7. doi:10.1016/j.bios.2004.07.021.

Sabater, S., Guasch, H., Ricart, M., Romaní, A., Vidal, G., Klünder, C. and Schmitt-Jansen, M. 2007. Monitoring the effect of chemicals on biological communities. The biofilm as an interface. Anal. Bioanal. Chem. 387: 1425–34. doi:10.1007/s00216-006-1051-8.

Siedenbiedel, F. and Tiller, J.C. 2012. Antimicrobial polymers in solution and on surfaces: overview and functional principles. Polymers (Basel). 4: 46–71. doi:10.3390/polym4010046.

Silley, P. and Forsythe, S. 1996. Impedance microbiology-a rapid change for microbiologists. J. Appl. Bacteriol. 233–243.

Stewart, G. 1899. The charges produced by the growth of bacteria in the molecular concentration and electrical conductivity of culture media. J. Exp. Med. 4.2 (1899): 235.

Stewart, P.S. 2003. New ways to stop biofilm infections. Lancet (London, England) 361: 97. doi:10.1016/S0140-6736(03)12245-3.

Stewart, P.S. and Franklin, M.J. 2008. Physiological heterogeneity in biofilms. Nat. Rev. Microbiol. 6: 199–210. doi:10.1038/nrmicro1838.

Suehiro, J., Hamada, R., Noutomi, D., Shutou, M. and Hara, M. 2003. Selective detection of viable bacteria using dielectrophoretic impedance measurement method. J. Electrostat. 57: 157–168. doi:10.1016/S0304-3886(02)00124-9.

Suehiro, J., Ohtsubo, A., Hatano, T. and Hara, M. 2006. Selective detection of bacteria by a dielectrophoretic impedance measurement method using an antibody-immobilized electrode chip. Sensors Actuators B Chem. 119: 319–326. doi:10.1016/j.snb.2005.12.027.

Sultana, S.T., Babauta, J.T. and Beyenal, H. 2015. Electrochemical biofilm control : a review. Biofouling 7014: 745–758. doi:10.1080/08927014.2015.1105222.

Tang, X., Flandre, D., Raskin, J.-P., Nizet, Y., Moreno-Hagelsieb, L., Pampin, R. and Francis, L.A. 2011. A new interdigitated array microelectrode-oxide-silicon sensor with label-free, high sensitivity and specificity for fast bacteria detection. Sensors Actuators B Chem. 156: 578–587. doi:10.1016/j.snb.2011.02.002.

UR, A. and Brown, D.F.J. 1975. Impedance monitoring of bacterial activity. Med. Microbiol. 8: 19–28.

Varshney, M., Li, Y. 2007. Interdigitated array microelectrode based impedance biosensor coupled with magnetic nanoparticle-antibody conjugates for detection of *Escherichia coli* O157:H7 in food samples. Biosens. Bioelectron. 22: 2408–14. doi:10.1016/j.bios.2006.08.030.

Varshney, M., Li, Y., Srinivasan, B. and Tung, S. 2007. A label-free, microfluidics and interdigitated array microelectrode-based impedance biosensor in combination with nanoparticles immunoseparation for detection of *Escherichia coli* O157:H7 in food samples. Sensors Actuators B Chem. 128: 99–107. doi:10.1016/j.snb.2007.03.045.

Varshney, M. and Li, Y. 2009. Interdigitated array microelectrodes based impedance biosensors for detection of bacterial cells. Biosens. Bioelectron. 24: 2951–60. doi:10.1016/j.bios.2008.10.001.

Wang, H., Mahdavi, A., Tirrell, A. and Hajimiri, A. 2012. Lab on a Chip PAPER A magnetic cell-based sensor. Lab Chip 12: 4465–4471. doi:10.1039/c2lc40392g.

Wang, Y., Ye, Z. and Ying, Y. 2012. New trends in impedimetric biosensors for the detection of foodborne pathogenic bacteria. Sensors 12: 3449–3471. doi:10.3390/s120303449.

Williams, D.L. and Bloebaum, R.D. 2010. Observing the biofilm matrix of Staphylococcus epidermidis ATCC 35984 grown using the CDC biofilm reactor. Microsc. Microanal. 16: 143–52. doi:10.1017/S143192760999136X.

Wilson, D. and Moragan, R. 2010. Medical device.

Yang, L., Ruan, C. and Li, Y. 2003. Detection of viable *Salmonella typhimurium* by impedance measurement of electrode capacitance and medium resistance. Biosens. Bioelectron. 19: 495–502. doi:10.1016/S0956-5663(03)00229-X.

Yang, L. 2008. Electrical impedance spectroscopy for detection of bacterial cells in suspensions using interdigitated microelectrodes. Talanta 74: 1621–9. doi:10.1016/j.talanta.2007.10.018.

Yang, L. and Bashir, R. 2008. Electrical/electrochemical impedance for rapid detection of foodborne pathogenic bacteria. Biotechnol. Adv. 26: 135–50. doi:10.1016/j.biotechadv.2007.10.003.

7

Rapid Microbial Water Quality Measurement by Automated Determination of the Fecal Indicator Bacterium *Escherichia coli*

A Review

Maximilian Lackner

1. Introduction

It is estimated that infectious diseases account for almost 40 per cent of all deaths globally, with an even higher share in developing countries (Abdel-Hamid et al. 1999). Cases of food-borne-illness, which result 91 per cent from bacteria, are on the order of 81 million cases per year—alone in the USA. In the water treatment and supply industry, the transmission of diseases is based on inappropriate treatment methods, failure in operation and supervision, or shortcomings in quality monitoring (Blasco and Pico 2009).

Institute of Advanced Engineering Technologies, University of Applied Sciences FH Technikum Wien, Höchstädtplatz 6, 1200 Vienna, Austria.
E-mail: maximilian.lackner@technikum-wien.at

It can further be argued that all water-borne diseases are preventable, by appropriate monitoring and timely corrective actions (WHO 2006). Water-borne disease outbreaks have been increasing in the last years (Araujo et al. 2014), and with cities becoming larger, it can be expected that severity of microbial water contamination incidents will rise as well.

2. Microbial Water Quality Determination

Due to the large number and diversity of potentially harmful microorganisms, it is useful to rely on the detection of suitable indicator bacteria to infer microbial water quality. Thermotolerant coliforms (ThCs) and enterococci have been deployed as fecal indicator bacteria (FIB) (Fiksdal and Tryland 2008). FIB have been used successfully for more than a century as surrogates for human enteric pathogens.

The coliform group includes a broad range of species, whether or not they belong to the Enterobacteriaceae family. Most definitions of coliforms are based on common biochemical traits. In "Standard Methods for the Examination of Water and Wastewater" (APHA 1998), coliform group members are described *as "all aerobic and facultative anaerobic, Gram-negative, non-spore-forming, rod-shaped bacteria that ferment lactose with gas and acid formation within 48 hours at 35°C* (multiple-tube fermentation technique (Edberg et al. 1988)). By contrast, ThCs are able to ferment lactose at higher temperatures, 44–45°C (WHO 2008). However, little correlation between ThC exposure and the incidence of acute gastroenteritis in beachgoers was found (Hamilton et al. 2005), but a significant correlation was detected between the illness rates in freshwater swimmers and their exposure to *Escherichia coli (E. coli)* (Hamilton et al. 2005).

E. coli can be differentiated from the other thermotolerant coliforms by their ability to produce indole from tryptophan or by production of the enzyme β-glucuronidase (WHO 2008). β-glucuronidase (GUS, EC 3.2.1.31) is an acid hydrolase. In humans, this lysosomal enzyme, which is found in bile, catalyzes the hydrolysis of β-D-glucuronidic bonds of several glucuronide esters (Kurtin and Schwesinger 1985). Generally, it degrades conjugated glucuronides.

E. coli is the most common intestinal microorganism of warm-blooded animals and humans. A conversion factor between ThC and *E. coli* of 0.63 was suggested by the US Environmental Protection Agency EPA (Health effects criteria for fresh recreational waters 1984), with findings from the pertinent literature, ranging from 0.57 to 0.81 (Hamilton et al. 2005).

Advantages of using *E. coli* as indicator bacteria for fecal contamination are

- Presence in large number in human and animal feces (10^9 colonies/g)
- Longer survival rate than many pathogens
- No regrowth under most environmental conditions
- Simple detection and quantification possible

E. coli are by themselves mostly not pathogenic, except for certain toxin-producing strains, such as O157:H7 (EHEC) that can cause diarrhea, urinary tract infections, inflammations and peritonitis, particularly in immune-suppressed people, such as children and the elderly (Zhang et al. 2009). However, since they are encountered together with other bacteria and toxins, their presence can serve as a warning signal.

According to the WHO, *E. coli* is considered the most suitable index of fecal contamination and it is used for surveillance of drinking water quality, including the use as disinfection indicator (WHO 2008).

Figure 1 provides an example of traditionally measured *E. coli* levels in a Californian river for illustration purposes.

At it can be seen from Fig. 1, there are highly and less contaminated areas, with different risks associated. Therefore, reliable *E. coli* detection can spot human health risks arising from water so that mitigating actions can

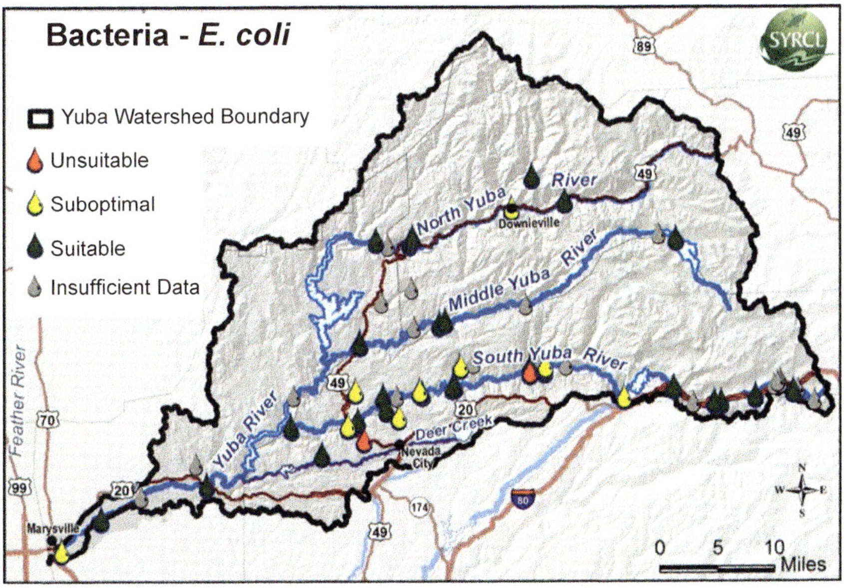

Figure 1. Monitoring sites where *E. coli* levels at the Yuba river watershed have been found to be unsuitable, suboptimal, or suitable, unless insufficient data exists. Source: US Environmental Protection Agency.

be implemented. This chapter concentrates on techniques for the detection of *E. coli* as proxy for fecal contamination.

2.1 Quantification of microbial water quality

2.1.1 Conventional methods

Traditional methods to detect *E. coli* are based on culturing sampled bacteria on growth media. The standard ISO 6222:1999 *"Water quality—Enumeration of culturable micro-organisms"*, which was reviewed and confirmed in 2010 (International Organization for Standardization 1999) detects total viable bacteria by the enumeration of colonies on plate count agar (PCA) after incubation at 36°C and 22°C for 2 and 3 days, respectively. PCA is a disk-shaped, nutrient-rich medium on which bacterial colonies can be seen and counted by the naked eye. Results are reported in CFU/100 ml (CFU = colony forming units).

Ideally, each viable cell would grow into a separate colony in the ISO 6222 test setup. However, in Nature, many bacterial cells are encountered as 'agglomerates', such as chains or lumps, hence it is more practical to speak about CFU when interpreting the PCA plates. The unit CFU will, for this reason, undercount the number of living cells in the actual sample.

For drinking water, the limit for *E. coli* is 0 CFU/100 ml (WHO 2008), whereas for recreational waters, a typical recommendation is <235 CFU/100 ml (http://www.wibeaches.us/apex/f?p=181:2). Sampling frequencies are, e.g., set by the TCR rule of EPA (example of the US) (http://www.epa.gov/ogwdw/disinfection/tcr/pdfs/qrg_tcr_v10.pdf) (Total Coliform Rule). For bathing beaches in the European community (EC), there is a guideline compliance limit of 100 FC/100 ml and a maximum allowable concentration of 2000 FC per 100 ml (European Community Council Directive 1975[1]) with FC being fecal coliforms, although meanwhile, *E. coli* was found to be a better indicator (Lebaron et al. 2005). See also the discussion above, *E. coli* represent 80–95 per cent of FC counts (Lebaron et al. 2005).

A similar approach is taken by the food industry, where *E. coli* O157 is one of the major four pathogens, and where the limit of detection by classical plating methods is approx. 4 CFU/ml for liquid foods and approx. 40 CFU/g for solid foods (Jasson et al. 2010).

Apart from the long-time duration needed to perform this test, there is a significant disadvantage: Only culturable bacteria are detected and these constitute only a fraction of all viable bacteria: Bacteria in Nature are

[1] Revised Bathing Europe Water Directive 2006/7/EC.

subjected to various stress conditions, such as UV light, nutrient deficiency and chemicals, so that when they are placed on a nutrient-rich PCA medium, they will show an extended lag-phase before starting to multiply. After 2–3 days, they are still viable but non-culturable (VBNC, active but non-culturable, ABNC) (Oliver 2005).

Philippe Lebaron (Lebaron et al. 2005) writes with reference to these culture-based techniques: *"E. coli concentrations may be underestimated at low concentrations when determined"*. Karina Yew-Hoong Gin (Gin and Goh 2013) raises the following concern: *"... that VBNC bacteria are being routinely ignored and thus, underestimated in the environment. Potentially this could pose a health risk to the public, especially for swimmers and other contact recreational activities in surface waters."*

In fact, depending on the environmental system,only a small portion (0.1–15 per cent) of the total bacterial population can be seen and enumerated by cultivation-based methods (Amann et al. 1990). However, detection of bacteria in VBNC state is important for two reasons:

- First, VBNC cells can still produce virulence proteins (Alleron et al. 2013).
- Second, when used as indicator bacteria, they still signal the presence of potential pathogens.

An alternative to classic plate counting is to perform culturing and counting on oligotrophic R2A agar (van der Linde et al. 1999, Reasoner et al. 1979). After 11 days of incubation (Lepeuple et al. 2004), which is too time-consuming for most practical applications though.

Hence, with state-of-the-art in *E. coli* determination having a time lag of 1–3 days, industry urgently demands a rapid detection system that yields result on all viable bacteria (total viable count, TVC) or FIB (*E. coli*) at least within 24 hours. Annie Rompré (Rompre et al. 2002) writes, *" The need for more rapid, sensitive and specific tests is essential in the water industry."* That demand is being met by advanced methods as described in the next section.

2.1.2 Advanced methods

Several 'fast' techniques have been developed over the last few years to detect *E. coli* and other bacteria more rapidly. Polymerase chain reaction (PCR), *in situ* hybridization techniques (FISH, fluorescent *in situ* hybridization) and immunoassays can be used to speed up measurement of *E. coli* (Zhang et al. 2009) down to several hours. Immunoassays use a specific antibody in combination with labels. A classic example is ELISA (enzyme-linked immunosorbent assay), where a secondary antibody is conjugated to an enzyme that forms a colored precipitate, another one ELFA (enzyme linked fluorescent assay).

A typical detection limit for PCR is 70 CFU/ml within 70 min (Lopez-Roldan et al. 2013). PCR can be used to identify sources of fecal contamination (Carlos et al. 2012) (microbial source tracking, MST). However, there are two practical drawbacks of PCR for *E. coli* detection (Heijnen and Medema 2009):

- Contamination of PCR reagents with trace amounts of *E. coli* DNA reduces the detection limit.
- PCR detectable DNA appears to be very stable after cell-death, resulting in detection of both viable and dead cells.

An approach similar to PCR is the nucleic acid sequence-based amplification (NASBA) method, which can be used for rapid detection of viable *E. coli* in water samples (Reasoner et al. 1979, Heijnen and Medema 2009). The sensitivity of the NASBA (Heijnen and Medema 2009) assay was found to be comparable with the culture method and approached a sensitivity of 1 CFU/100 ml.

Results were obtained in 3–4 hours (Lopez-Roldan et al. 2013).

Ihab Abdel-Hamid (Abdel-Hamid et al. 1999) reports a flow injection immunoassay system to detect 50 cells/ml of *E. coli* within 35 minutes based on the nylon membranes used as support for anti-*E. coli* antibodies.

Another approach reported by Xinai Zhang (Zhang et al. 2009) is an electrochemical immunoassay of *E. coli* using anodic stripping voltammetry based on Cu@Au nanoparticles as antibody labels, where detection limits of 30 CFU/ml within 2 hours were reached, which could be lowered to 3 CFU/10 ml by a pre-enrichment step.

In the paper of Lepeuple et al. (Lepeuple et al. 2004), a technique to detect viable *E. coli* based on their membrane esterase activity is discussed. The system (Chem Chrome V6, TVC Bioburden) uses direct fluorescent labelling of viable microorganisms in conjunction with an automated laser scanning and enumeration system (cytometry). Labelling is performed with a non-fluorescent fluorescein derivative specifically hydrolyzed by the enzymes of the target microorganisms (here: membrane esterase of *E. coli*), where only viable cells with active enzymes and integer membranes can retain the fluorescent product.

The TVC assay was compared to growth on R2A agar and labelling with CTC (cyano-ditolyl-tetrazolium chloride), which detects bacteria with a functioning respiratory chain (Cervantes et al. 1997). These methods yield the number of bacteria per ml or per 100 ml.

Another method dubbed 'fast' is IDEXX™ (Colilert18 (ISO 9308-2:2012), where MPN (most probable numbers) of bacteria per volume are obtained after typically 18 hours (http://www.wibeaches.us/apex/f?p=181:2).

In the IDEXX™ Colilert™ system, which is used for the simultaneous detection and enumeration of total coliforms and *E. coli* in water, two chromogenic nutrient indicators, i.e., ortho-nitrophenyl-β-D-galactopyranoside (ONPG) and 4-methylumbelliferyl-β-D-glucuronide (MUG) are used. As coliforms grow, they use β-galactosidase to metabolize ONPG and change it from colorless to yellow. *E. coli* deploy β-glucuronidase to metabolize MUG and create fluorescence (Jasson et al. 2010, Lopez-Roldan et al. 2013). Determining MPN in tubes (O'Toole and Chiang 1999), microtiter plates and presence/absence (P/A) (http://www.hardydiagnostics.com/articles/colitag-flier.pdf) tests are common (WHO 2008).

Gardner (Gardner et al. 1998) discusses an electronic nose for *E. coli* detection. Gases formed in the head space of bacterial cultures grown in a standard nutrient medium are analyzed; however, time is needed. A key question around such advanced methods is that of comparability of results, both towards classic PCA testing and towards other methods.

A correlation between PCR quantification and culture-based methods has been achieved and reported in the publication of Lopez-Roldan et al. (Lopez-Roldan et al. 2013).

Figure 2 shows an excerpt from Lepeuple's paper (Lepeuple et al. 2004) where the results from these three advanced enumeration methods are depicted as bacteria/ml.

What can be seen and inferred from Fig. 2 is that the parameter 'bacteria per ml' is not robust either, since the 3 methods yield different

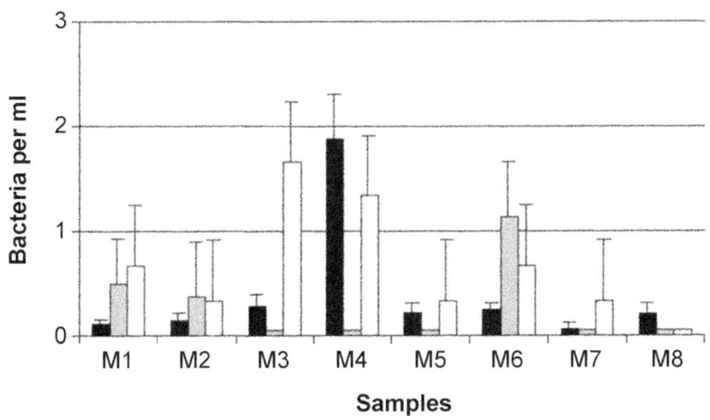

Figure 2. Enumeration of total bacteria in drinking water samples. Three enumerating methods were used TVC (black), CTC (grey) and R2A (white).

results. Lepeuple concludes (Lepeuple et al. 2004): *"Indeed, in the case of GAC [granular activated carbon]-filtered water samples, TVC counts appeared to be much lower than CTC counts by up to 2 log 10 orders of magnitude. In the case of drinking water samples, TVC counts were not significantly different from CTC counts. These results are in contradiction with the literature where TVC counts appeared to be significantly higher than CTC counts for mineral and potable water samples only. In addition, it is difficult to compare the two sets of data as the two dyes TVC and CTC have different target sites."*

Another issue generally associated with advanced methods for fast *E. coli* detection is complexity. Most of the above-mentioned techniques require a laboratory with well-trained staff. Enzymatic assays seem to have the best potential for automation and will be discussed more deeply below.

2.2 Enzymatic assays

Enzymes are very specific biological catalysts, which can be used on chromogenic and fluorogenic substrates to produce color and fluorescence, respectively, upon cleavage by a specific enzyme. The specificity of enzymes for a particular substrate is of high importance in biological systems.

The enzyme β–glucuronidase is a characteristic *E. coli* enzyme found in >95 per cent of all *E. coli* (Hartman 1989, Huang 2008). It can cleave β-D-glucoronidic bonds. Its reactivity can be used to design a fast test for *E. coli* specific activity. For total coliforms, the enzyme β-D-galactosidase can be used for the design of such a fast test method (Rompre et al. 2002).

Chromogenic substrates, composed of glucuronide and a chromogen, have been used in several commercial products to indicate the presence of *E. coli*, amongst them m-ColiBlue24™ (Hach Company), where the reaction of indoxyl-β-D-glucuronide with the enzyme β-glucuronidase produces a color change. Other media, such as Colilert™ (IDEXX Labs) and nutrient agar with 4-methylumbelliferyl-β-D-glucuronide (MUG) (Difco Company), release the fluorescent product 4-methylumbelliferone from the reaction of β-glucuronidase with MUG (Hamilton et al. 2005).

In the publication of Lebaron et al. (Lebaron et al. 2005), the potential application of the β-D-glucuronidase activity measurement for the routine detection and quantification of *E. coli* in marine bathing waters was investigated. GUS activity was measured as the rate of hydrolysis of 4-methylumbelliferyl-β-D-glucuronide. Culturable *E. coli* were quantified by the most probable number (MPN) microplate method. A significant correlation was found between the log of GUS activity and the log of culturable *E. coli*. It was concluded that *"the GUS activity is an operational, reproducible, simple, very rapid and low-cost method for the real-time enumeration of E. coli in bathing waters and should be preferred to the microplate method"* (Lebaron et al. 2005).

Dustin Starkey (Starkey et al.) presented a fluorogenic assay for β-glucuronidase using microchip-based capillary electrophoresis with the target of combinatorial drug screening. Hydrolysis with β-glucuronidase of the conjugated glucuronide, fluorescein mono-β-D-glucuronide (FMG), liberated the fluorescent product, fluorescein.

Another novel instrument based on a CMOS array photo sensor is discussed in the publication of Su-Hua Huang et al. (Huang 2008).

Fluorescence assays of β-glucuronidase activity have used 4-methylumbelliferyl-β-D-glucuronide (MUG). HPLC (high performance liquid chromatography) and CE (capillary electrophoresis) were used in three of these to separate and monitor the hydrolysis product, 4-methylumbelliferone, from the MUG substrate. Fluorescein mono-β-D-glucuronide (FMG) is another fluorescent substrate of β-glucuronidase that can be used in a separation-based assay. This conjugate is hydrolyzed by β-glucuronidase, liberating fluorescein. Fluorescein has a larger extinction coefficient than 4-methylumbelliferone, a larger quantum yield and a lower sensitivity to photochemical assays decomposition, making FMG an attractive choice (see Fig. 3).

The second commonly used reaction is depicted in Fig. 4 below.

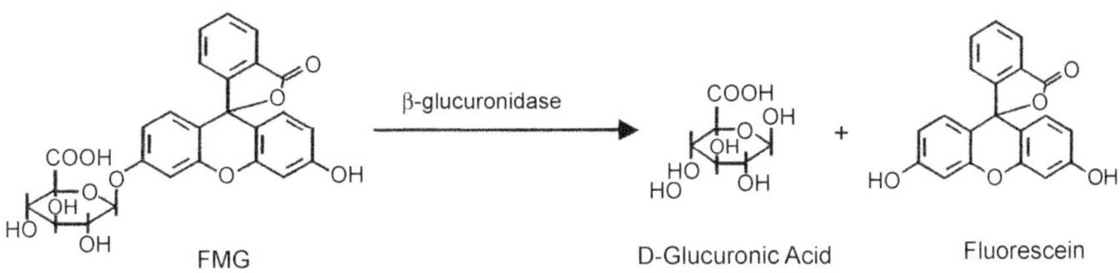

Figure 3. Hydrolysis of FMG by β-glucuronidase yielding fluorescein and D-glucuronic acid (Source: Starkey et al.).

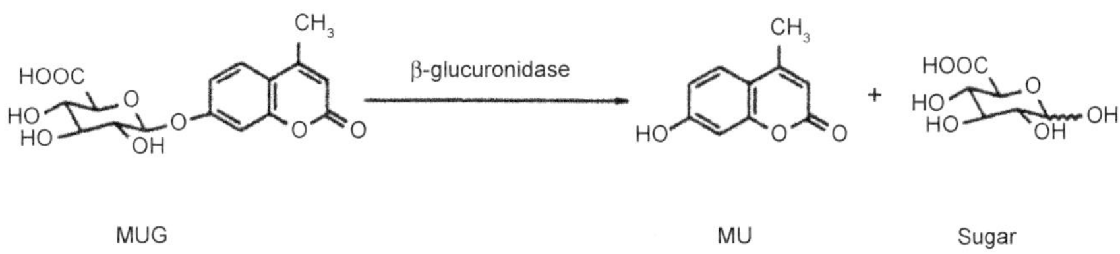

Figure 4. Conversion of substrate MUG to the products MU and D-glucuronic acid through the action of β-glucuronidase (Source: Starkey et al.).

Figure 5 shows the fluorescence emission spectra of MU and MUG, reproduced from Fior et al. (2009).

As Fig. 5 shows, the cleavage product MU as a result of beta-glucuronidase activity can be detected in the near UV.

To detect the presence of β-D-glucuronidase in *E. coli*, the following chromogenic substrates were also described in the literature (Rompre et al. 2002): indoxyl-β-D-glucuronide (IBDG), the phenolphthalein-mono-β-D-glucuronide complex and 5-bromo-4-chloro-3-indolyl-β-D-glucuronide. However, most frequently, the fluorogenic substrate 4-methylumbelliferyl-β-D-glucuronide (MUGlu) was used (Rompre et al. 2002).

Chromogenic substrates such as o-nitrophenyl-β-D-galactopyranoside (ONPG), p-nitrophenyl-β-D-galactopyranoside(PNPG), 6-bromo-2-naphtyl-β-D-galactopyranoside and 5-bromo-4-chloro-3-indolyl-β-D-galactopyranoside were used to detect the presence of β-D-galactosidase produced by coliforms, as well as the fluorogenic substrate 4-methylumbelliferyl-β-D-galactopyranoside (MUGal) (Rompre et al. 2002).

Other researchers used polycyclic aromatic hydrocarbon (PAH) conjugated substrates, for example, pyrene-galactopyranoside and anthracene-glucuronide. The enzymatic degradation, via β-glucuronidase and β-galactosidase activity, of these conjugated substrates releases hydroxylated PAH products (for example, hydroxypyrene or hydroxyanthracene) that can be detected (Wildeboer et al. 2010). Several studies have focused in the last few years on evaluating the indicator applicability of automated on-site measurements of GUS in ground- and surface waters (Ryzinska-Paier et al. 2014, Stadler et al. 2016).

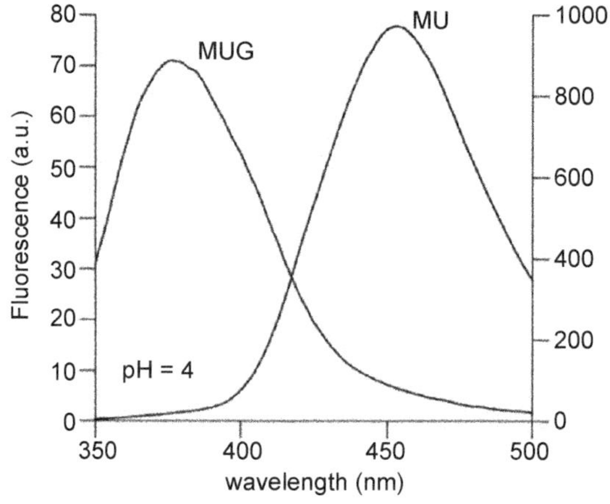

Figure 5. MU and MUG fluorescence emission spectra. 4 mM MU and MUG solutions in acetate buffer pH 4 have been excited at 323 and 318 nm, respectively. Left y-axis refers to MUG, right to MU (Source: (Fior et al. 2009)).

The fluorescence assay is stable and robust. It works on freshwater and seawater alike. Beta-glucuronidase is inhibited by glucaro-1,4-lactone (Kurtin and Schwesinger 1985), which is not present in typical natural samples. Interference can come from humic substances (Wildeboer et al. 2010). Figure 6 shows factors that can contribute to a measured signal of enzyme activity.

Some studies have proposed to estimate the β-D-glucuronidase activity of *E. coli* in rapid assays performed without any cultivation step as a surrogate of *E. coli*. Good correlations in log-log plot were generally found in natural waters between GUS activity and FC or *E. coli* levels (Farnleitner et al. 2002) see Fig. 7 below.

The detection limit of the study underlying Fig. 7 was 15 *E. coli*/100 ml.

Figure 8 below shows a comparison of a traditional lab measurement (in CFU/ml) vs. an instrument measurement based on enzyme activity (measured in GUD (GUS) activity). Fairly good correlation can be seen.

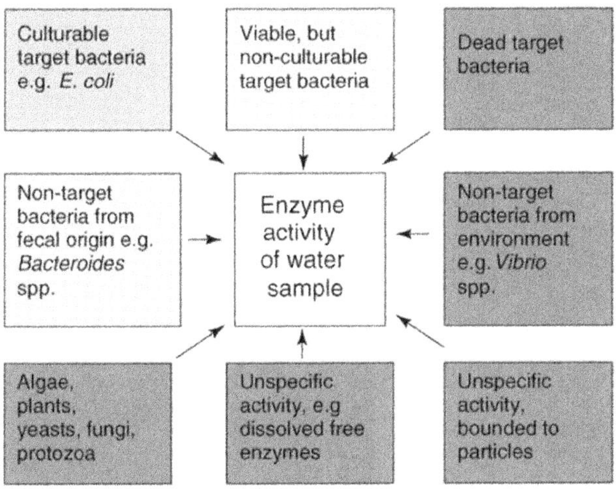

Figure 6. Overview of fractions that may contribute to measured enzyme activity in rapid enzyme assays for detection of fecal water contamination. Specific enzyme activities measured in field trials are often correlated to the number of culturable target bacteria (green box), and for log–log plots good correlations are generally obtained. Interference/contribution of viable but non-culturable target bacteria, as well as viable non-target bacteria from fecal origin (yellow boxes), can be acceptable in an alternative rapid enzyme assay for detection of fecal contamination. Interference from enzyme activity of non-fecal origin (red boxes) should be limited because of its lack of association with a potential health risk. The significance/size of the different fractions may vary depending on type of environmental water (fresh, marine) and environmental conditions (light, turbidity, nutrients, fecal contamination) (Source: (Fiksdal and Tryland 2008)).

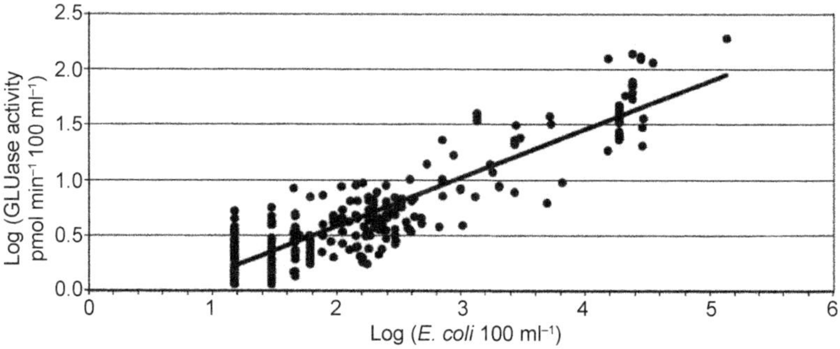

Figure 7. Log–log linear regression between GUS activity and *E. coli* concentrations in seawater samples: Log (GUS act. pmoles min^{-1} 100 ml^{-1}) = 0.436 Log (*E. coli* 100 ml^{-1}) 0.818 ($r^2 = 0.81$, $n = 256$, $p < 0.001$). Source: Lebaron et al. (Lebaron et al. 2005).

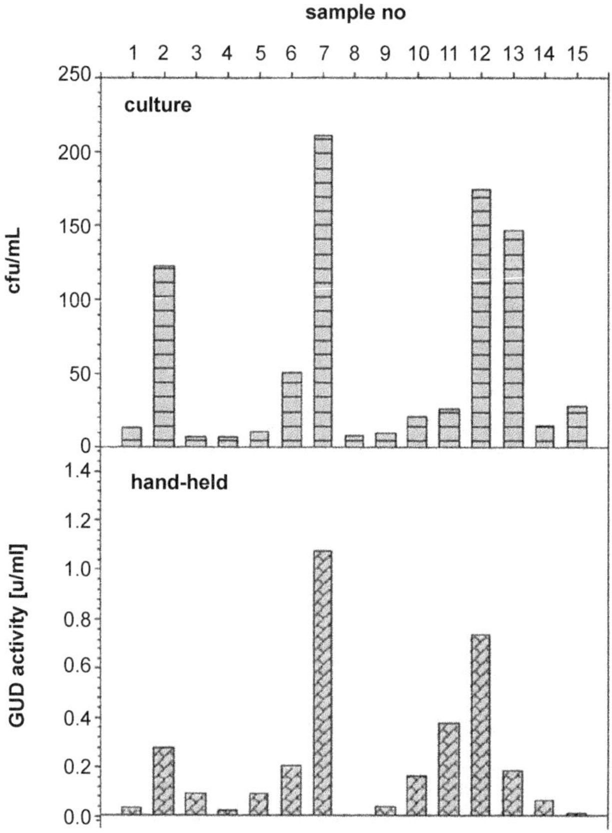

Figure 8. Comparison of the results obtained with the three methods for analyzing *E. coli* in River Thames water. 15 samples were analysed with the direct assay using the hand-held detector (middle panel), by filtration and subsequent cultivation and counting of *E. coli* colony forming units (CFU) per mL (top panel) (Source: Wildeboer et al. (2010)).

2.3 Chemiluminescence

An alternative to enzymatic assays is chemiluminescence. Bukh et al. (Bukh and Roslev 2010) describe the hydrolysis of chemiluminescent 1,2-dioxetanes by *E. coli*. Chemiluminescence yielded greater sensitivity than fluorescence. Combined with membrane filtration, it showed potential for early-warning detection of microbial contaminations in drinking water (4–6 hours) (Bukh and Roslev 2010). Yet another approach is based on ATP bioluminescence (Lebaron et al. 2005).

It has been used widely in the food and beverage industries (Lopez-Roldan et al. 2013), see Fig. 9 below.

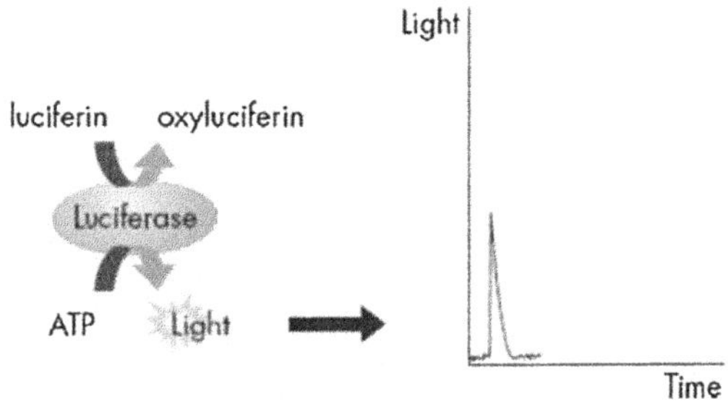

Figure 9. ATP luminescence measurement (Source: Lopez-Roldan et al. (Lopez-Roldan et al. 2013)).

2.4 Flow cytometry

Another method to determine bacteria in water is flow cytometry. Flow cytometry of *E. coli* has been proposed to track disinfection efficiency (Riccardo Bigoni et al. 2014, Simon Gillespie et al. 2014) and for process monitoring (Scheper et al. 1987). A major disadvantage is that essentially all solids in water are detected.

2.5 Other techniques

Additional techniques, especially molecular detection schemes, are discussed in (Douterelo et al. 2014), see Fig. 10.

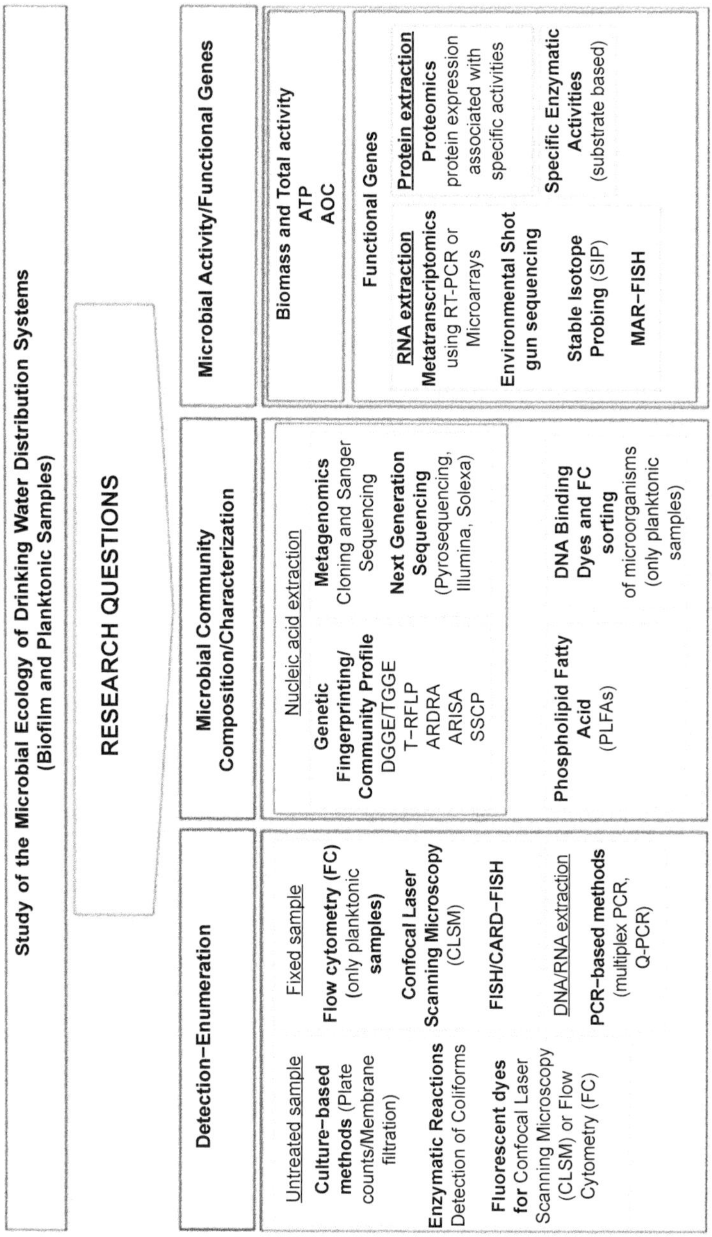

Figure 10. Scheme showing the different techniques available to characterize microbial communities in drinking water distribution systems. Source: Douterelo et al. (Douterelo et al. 2014).

3. Comparison of Available Instruments

In Lopez-Roldan's paper (Lopez-Roldan et al. 2013), several techniques and commercial instruments for bacterial contamination measurements of water samples are reviewed (see Table 1).

As Table 1 shows, the results are reported in CFU/ml and cells/ml. The authors of the table write: *"Few studies have reported dealing with validation of these instruments. Some are very new and, in a remarkable number of cases, results of the tests are confidential or performed by private companies or other institutions with no interest in publication of results* (Lopez-Roldan et al. 2013).

Annie Rompré (Rompre et al. 2002) shows a similar comparison of detection limits by PCR, quoted in cells, CFU and cells/100 ml.

Vicky Jasson has presented the review of alternative microbial methods alongside the European and US approach for their validation (food industry, ISO 16140) (Jasson et al. 2010).

Table 1. Comparison of Commercially Available Techniques for *E. coli* Detection (LoD = Limit of Detection).

Type of technology	Parameters analyzed in water samples	LoD	Analysis Time
Light scattering	*Cryptosporidium*	1 oocysts/mL	60 min
	E. coli	1,000 CFU/mL	Continuous
ATP luminescence	Total microbial biomass	200 CFU/mL	5 min
Immunoassays	*Bacillus anthracis*	100,000 CFU/mL	15 min
	Legionella pneumophila	800 CFU/mL	180 min
Polymerase chain reaction (PCR)	*E. coli, Enterococci*	15 CFU/100 mL	3 h
	Legionella pneumophila	100 genomes units/L	3 h
Enzyme fluorescence	*E. coli*, total coliforms	50 CFU/100 mL	1 h
FISH	*E. coli*	100-100 CFU/mL	2 h
		10 CFU/100 mL	10 h
Molecularly-imprinted polymers	*Bacillus*	not specified	1 day (plus bead synthesis)
Electrochemiluminescence (ECL)	*E. coli, Salmonella*	1,000 cells/mL	1 h
Raman spectroscopy	*E. coli*	1,000 cells/mL	
Dye-loaded microspheres	*E. coli*	1,000 cells/mL	1 h

Stadler et al. compared different prototypes for the automated on-site GUS measurements (Stadler et al. 2016).

4. A New Parameter: MFU/100 ml

In order to solve the uncertainty issues introduced by measuring in number of cells/100 ml (which can be dead, viable or culturable) and CFU/100 ml (where a target of 0 is also not an ideal value to be measured), a novel, more robust parameter was introduced by Vogl et al. (Vogl et al. 2013). The unit MFU (modified Fishman units) is defined as enzymatic activity as follows:

$$\text{Phenolphthalein Glucuronide} + H_2O \xrightarrow{\beta\text{-Glucuronidase}} \text{D-Glucuronate} + \text{Phenolphthalein}$$

Conditions: Temperature = 37°C, pH = 6.8.

1 MFU will liberate 1.0 µg of phenolphthalein from phenolphthalein glucuronide per hour at pH 6.8 at 37°C. For details on MFU, see (Fishman and Bernfeld 1955, Combie et al. 1982, Fishman 1974, (http://www.sigmaaldrich.com/technical-documents/protocols/biology/enzymatic-assay-of-b-glucuronidase-from-ecoli.html#sthash.02JpaIAX.dpuf)).

Vogl proposed to tie the value 'MFU' to the volume of 100 ml, hence yielding the parameter 'MFU/100 ml' (Vogl et al. 2013). Based on his definition of MFU, one 'modified Fishman unit/100 ml' will liberate 1.0 µg of phenolphthalein from phenolphthalein glucuronide per 100 ml per hour at pH 6.8 at 37°C. By relating MFU to a certain volume, the degree of bacterial contamination can be set into perspective with currently obtained results in CFU per the same volume, compare also Fig. 8. So instead of determining the number of cells or colonies of bacteria, the authors propose to measure their enzymatic activity, to which culturable and viable cells, but no dead cells, contribute. A major advantage of 'MFU/100 ml' is that it is independent on the instrument being used. Figure 11 shows a linear relationship between enzyme units added to a sample and the observed activity.

MFU/100 ml can be measured faster than CFU/100 ml, hence allowing critical decisions to be taken significantly faster. There is no under prediction of bacterial contamination, as both culturable and viable but non-culturable bacteria are detected. In contast to PCR, no dead bacteria are included.

Hamilton (Hamilton et al. 2005) concludes that *"more water quality violations will occur when enzyme-specific media are used for testing than if conventional culture media are used."*

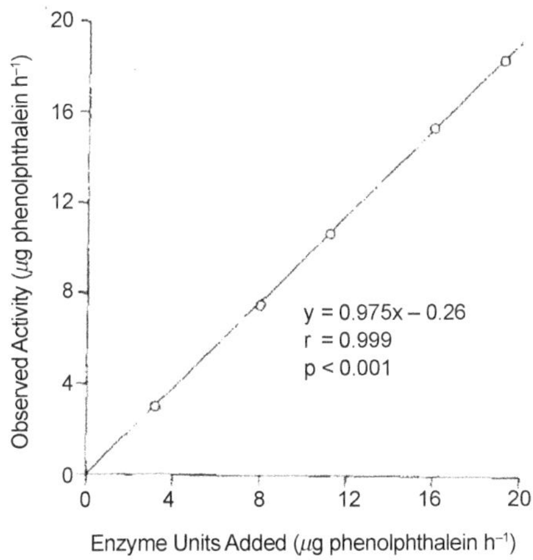

Figure 11. Linear relationship between measured Beta-glucuronidase activity in hepatic bile samples vs. the units of exogeneous enzyme added prior to extraction (Source: Kurtin et al. (Kurtin and Schwesinger 1985)).

Hence a better threshold than '0' needs to be established.

Figure 12 shows a direct comparison of measurements of a freshwater river sample in MFU/100 ml vs. CFU/100 ml over a period of two weeks.

As can be seen above, culturability of *E. coli* (CFU/100 ml) is rapidly decreasing in water. Time for 90 per cent reduction, t_{90}[d] is less than 3 days. After 6 days, *E. coli* bacteria are not detectable by traditional cultivation methods. Enzymatic activity (MFU/100 ml) remains significantly longer. Time for 90 per cent reduction, t_{90}[d] is more than 11 days.

The prolonged t_{90}[d] for fecal-associated enzymatic activity method allows to detect fecal traces even two weeks after contamination.

Note that there is no fixed conversion factor between CFU/100 ml and MFU/100 ml (1:21, 500 in Fig. 13), since the ratio of colony-forming *E. coli* to *E. coli* in VBNC-state is theoretically between 1:0 and 0:1. A detailed comparison of MFU/100 ml, GUS activity in pmol/min/100 ml and culture based *E. coli* analyses can be found in (Stadler et al. 2016).

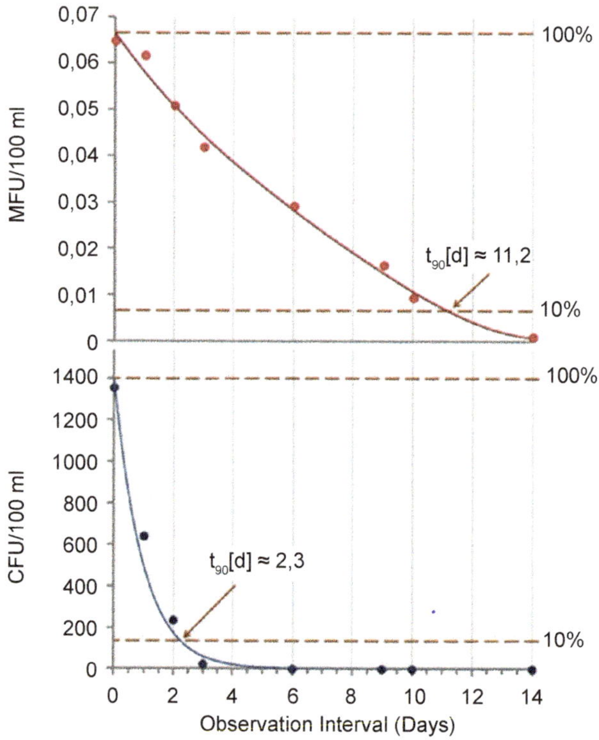

Figure 12. Fecal contamination of a water sample measured over 14 days. Legend: CFU/100 ml (Colony Forming Units in 100 ml water): amount of culturable bacteria (*E. coli*) able to grow and form colonies in special nutrition media; MFU/100 ml: specific enzymatic activity of β-Glucuronidase present in metabolically active *E. coli* (Source: Vogl et al. 2013).

5. Towards Online Monitoring of Water Quality

Already today, water utilities worldwide employ on-line monitoring tools and early warning systems at various stages for measuring physical properties and some chemical compounds. Real-time detection of chemical, biological, radiological, or nuclear (CBRN) contaminants is often required for planning and implementing mitigation measures to protect water supplies (Storey et al. 2011, Sanjay and Charles 2006).

So far, this has not been possible with microbiological parameters, due to the high time requirements for analysis.

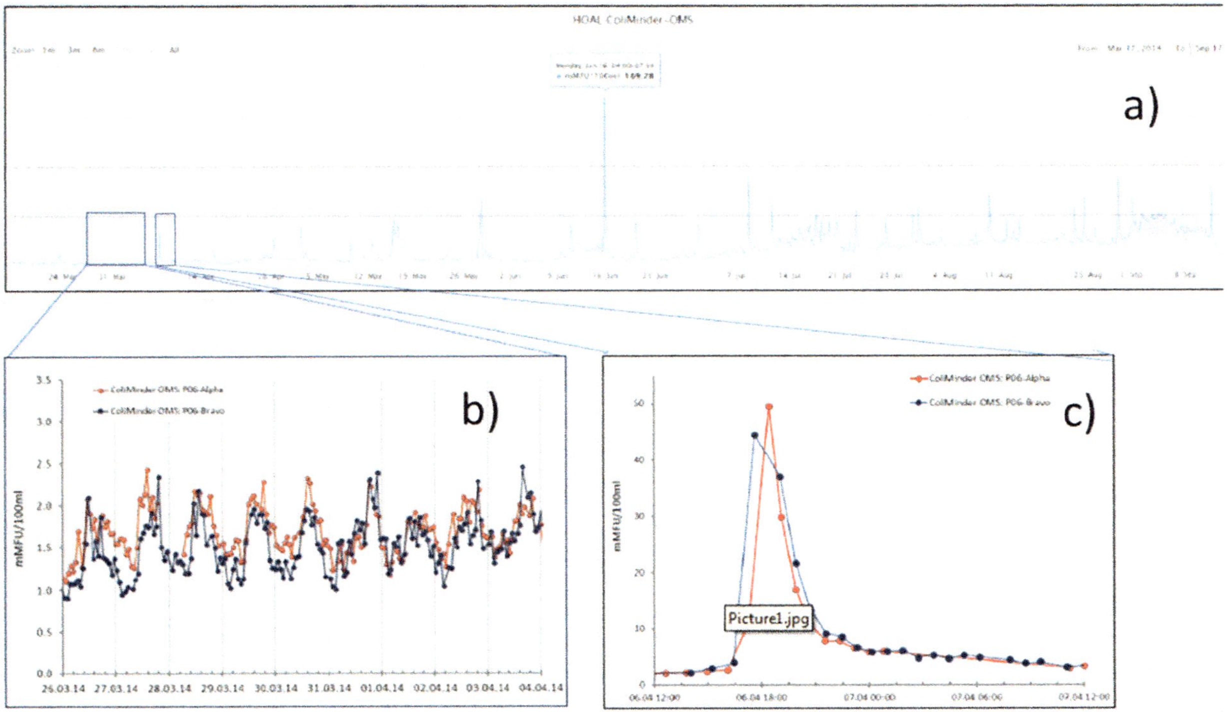

Figure 13. Surface water monitoring of an Austrian creek. (a) 6 month time period; (b) Diurnal fluctuations; (c) Probing of a rainfall event. The red and blue curves are 2 instruments which were gated intermittently (Source: Koschelnik et al. (Koschelnik et al. 2014)).

Quantitative Microbial Risk Assessment (QRMA) is a tool of assessing risk to the general population arising from microbial contamination, e.g., on beaches. QMRA (Haas et al. 1999) requires near-real-time data of microbiological water quality.

An example of near-real-time water quality monitoring is given below. Koschelnik et al. (Koschelnik et al. 2014) have measured the level of *E. coli* in an Austrian creek over a 6-month period with approx. 23 measurements per day (see Fig. 13 below).

Compared to the situation in Fig. 1, where a traditional method was used to obtain sparse data, the fast and automated technology deployed in Fig. 13 by Koschelnik et al. (Koschelnik et al. 2014) offers a time resolution of 30 minutes and below, which makes the technology suitable not only for surface water monitoring, but also for process control applications. A potential process control application could be, for instance, in disinfection (i.e., use of a minimum amount of chemicals or energy to achieve a desired degree of bacteria inactivation or selection of the best available raw water stream).

6. Summary and Conclusions

In this chapter, the measurement of microbiological water quality by detection of fecal indicator bacteria was reviewed. *E. coli* is suited best to signal the potential presence of pathogens. However, traditional methods are time-consuming and yield results which tend to underestimate the actual bacterial burden. Advanced methods, such as PCR, chemiluminescence (ATP) and enzymatic assays need between 15 minutes and 6 hours to yield a quantitative result, which can include bacteria in the VNBC-state; hence providing a higher level of security as to whether fecal contamination is present in a given water sample. Such fast methods have the potential to be used in monitoring networks and associated risk map generations (QMRA, quantitative microbiological risk assessment) and process control applications, e.g., for the determination of optimum disinfection levels so that energy and chemicals can be saved.

Near real time monitoring of *E. coli* has a huge potential, and novel instruments have come close to commercialization on a larger scale.

Keywords: Fecal contamination, process control, water quality, *E. coli*, MUG, GUS, drinking water safety, CFU (Colony Forming Units), VBNC (Viable But Non-Culturable), ABNC (Active But Non-Culturable), MFU (Modified Fishman Units)

References

Abdel-Hamid, I., Ivnitski, D., Atanasov, P. and Wilkins, E. 1999. Flow-injection immunosystem. Analytica Chimica Acta 399: 99–108.

Alleron, L., Khemiri, A., Koubar, M., Lacombe, C., Coquet, L., Cosette, P., Jouenne, T. and Frere, J. 2013. VBNC Legionella pneumophila cells are still able to produce virulence proteins. Water Res. 47: 6606–6617.

Amann, R.I., Krumholz, L. and Stahl, D.A. 1990. Fluorescent-oligonucleotide probing of whole cells for determinative, phylogenetic, and environmental studies in microbiology. J. Bacteriol. 172: 762–770.

American Public Health Association (APHA). 1998. Standard Methods for the Examination of Water and Wastewater, 23rd Edition, American Water Works Association, ISBN: 978-1625762405 (2017), http://www.standardmethods.org/.

Araujo, S., Henriques, I.S., Leandro, S.M., Alves, A., Pereira, A. and Correia, A. 2014. Gulls identified as major source of fecal pollution in coastal waters: a microbial source tracking study. Sci. Total Environ. 470-471: 84–91.

Blasco, C. and Pico, Y. 2009. Prospects for combining chemical and biological methods for integrated environmental assessment. Trends Anal. Chem. 28: 745–757.

Bukh, A.S. and Roslev, P. 2010. Characterization and validation of a chemiluminescent assay based on 1,2-dioxetanes for rapid detection of viable *Escherichia coli*. Appl. Microbiol. Biotechnol. 86: 1947–1957.

Carlos, C., Alexandrino, F., Stoppe, N.C., Sato, M.I. and Ottoboni, L.M. 2012. Use of *Escherichia coli* BOX-PCR fingerprints to identify sources of fecal contamination of water bodies in the State of Sao Paulo, Brazil. J. Environ. Manage. 93: 38–43.

Cervantes, P., Mennecart, V., Robert, C., de Roubin, M.R. and Joret, J.C. 1997. The microbiological quality of water (Ambleside, UK: Freshwater Biological Association), pp. 54–62.

Combie, J., Blake, J.W., Nugent, T.E. and Tobin, T. 1982. Morphine glucuronide hydrolysis: superiority of beta-glucuronidase from Patella vulgata. Clin. Chem. 28: 83–86.

Edberg, S.C., Allen, M.J. and Smith, D.B. 1988. National field evaluation of a defined substrate method for the simultaneous enumeration of total coliforms and *Escherichia coli* from drinking water: comparison with the standard multiple tube fermentation method. Appl. Environ. Microbiol. 54: 1595–1601.

Enzymatic Assay of β-Glucuronidase (EC 3.2.1.31) from *E. coli* (http://www.sigmaaldrich.com/technical-documents/protocols/biology/enzymatic-assay-of-b-glucuronidase-from-ecoli.html#sthash.02JpaIAX.dpuf).

Farnleitner, A.H., Hocke, L., Beiwl, C., Kavka, G.G. and Mach, R.L. 2002. Hydrolysis of 4-methylumbelliferyl-beta-D-glucuronide in differing sample fractions of river waters and its implication for the detection of fecal pollution. Water Res. 3: 975–981.

Fiksdal, L. and Tryland, I. 2008. Application of rapid enzyme assay techniques for monitoring of microbial water quality. Curr. Opin. Biotechnol. 19: 289–294.

Fior, S., Vianelli, A. and Gerola, P.D. 2009. A novel method for fluorometric continuous measurement of β-glucuronidase (GUS) activity using 4-methyl-umbelliferyl-β-d-glucuronide (MUG) as substrate. Plant Science 176: 130–135.

Fishman, W.H. 1974. Methods of Enzymatic Analysis (N.Y., USA: Academic Press), pp. 930–932.

Fishman, W.H. and Bernfeld, P. 1955. Meth. Enzymol., pp. 262–269.

Gardner, J.W., Craven, M., Dow, C. and Hines, E.L. 1998. The prediction of bacteria type and culture growth phase by an electronic nose with a multi-layer perceptron network. Measurement Science and Technology 9: 120.

Gin, K.Y. and Goh, S.G. 2013. Modeling the effect of light and salinity on viable but non-culturable (VBNC) Enterococcus. Water Res. 47: 3315–3328.

Haas, C., Rose, J. and Gerba, C. 1999. Quantitative Microbial Risk Assessment: John Wiley and Sons.

Hamilton, W.P., Kim, M. and Thackston, E.L. 2005. Comparison of commercially available *Escherichia coli* enumeration tests: implications for attaining water quality standards. Water. Res. 39: 4869–4878.

Hardy Diagnostics (http://www.hardydiagnostics.com/articles/colitag-flier.pdf).

Hartman, P.A. 1989. Rapid Methods and Automation in Microbiology and Immunology (Springer, Berlin), pp. 290–308.

Health effects criteria for fresh recreational waters 1984 (Washington, DC: USEPA. Office of Research and Development).

Heijnen, L. and Medema, G. 2009. Method for rapid detection of viable *Escherichia coli* in water using real-time NASBA. Water Res. 43: 3124–3132.

Huang, S.-H. 2008. Detection of *Escherichia coli* using CMOS array photo sensor-based enzyme biochip detection system. Sensors and Actuators B: Chemical. 133: 561–564.

Isabel Douterelo, Joby B. Boxall, Peter Deines, Raju Sekar, Katherine E. Fish and Catherine A. Biggs. Methodological approaches for studying the microbial ecology of drinking water distribution systems, Water Research, Volume 65, 15 November 2014, pp. 134–156.

ISO 1999. (Geneva, ISO: International Organization for Standardization). *Water Quality – Enumeration of Culturable Micro-organisms – Colony Count by Inoculation in a Nutrient Agar Culture Medium* 1999.

Jasson, V., Jacxsens, L., Luning, P., Rajkovic, A. and Uyttendaele, M. 2010. Alternative microbial methods: An overview and selection criteria. Food Microbiol. 27: 710–730.

Juri Koschelnik, Markus Epp, Wolfgang Vogl, Philip Stadler, Maximilian Lackner MFU/100 ml: New Measurement Parameter for Rapid Enzymatic Monitoring of Fecal-Associated Indicator Bacteria in Water, 2014 Water & Health Conference, UNC Water Institute, USA, 2014.

Kurtin, W.E. and Schwesinger, W.H. 1985. Assay of beta-glucuronidase in bile following ion-pair extraction of pigments and bile acids. Anal. Biochem. 147: 511–516.

Lebaron, P., Henry, A., Lepeuple, A.S., Pena, G. and Servais, P. 2005. An operational method for the real-time monitoring of *E. coli* numbers in bathing waters. Mar. Pollut. Bull. 50: 652–659.

Lepeuple, A.S., Gilouppe, S., Pierlot, E. and De Roubin, M.R. 2004. Rapid and automated detection of fluorescent total bacteria in water samples. Int. J. Food Microbiol. 92: 327–332.

Lopez-Roldan, R., Tusell, P., Cortina, J.L., Courtois, S. and Cortina, J.L. 2013. On-line bacteriological detection in water. TrAC Trends in Analytical Chemistry 44: 46–57.

O'Toole, D.K. and Chiang, M.M.P. 1999. The use of MUG supplement to detect *Escherichia coli* by the multiple tube method in marine waters of Hong Kong. Marine Pollution Bulletin 38: 921–924.

Oliver, J.D. 2005. The viable but nonculturable state in bacteria. J. Microbiol. 43SpecNo: 93–100.

Reasoner, D.J., Blannon, J.C. and Geldreich, E.E. 1979. Rapid seven-hour fecal coliform test. Appl. Environ. Microbiol. 38: 229–236.

Riccardo Bigoni, Stefan Kötzsch, Sabrina Sorlini and Thomas Egli. 2014. Solar water disinfection by a Parabolic Trough Concentrator (PTC): flow-cytometric analysis of bacterial inactivation. Journal of Cleaner Production, Volume 67, 15 March 2014, pp. 62–71.

Rompre, A., Servais, P., Baudart, J., de-Roubin, M.R. and Laurent, P. 2002. Detection and enumeration of coliforms in drinking water: current methods and emerging approaches. J. Microbiol. Methods 49: 31–54.

Ryzinska-Paier, G., Lendenfeld, T., Correa, K., Stadler, P., Blaschke, A.P., Mach, R.L., Stadler, H., Kirschner, A.K.T. and Farnleitner, A.H. 2014. A sensitive and robust method for automated on-line monitoring of enzymatic activities in water and water resources. Water Sci. Technol. J. Int. Assoc. Water Pollut. Res. 69: 1349–1358. doi:10.2166/wst.2014.032.

Sanjay, J. and Charles, M.R. 2006. An integrating framework for modeling and simulation for incident management. Journal of Homeland Security and Emergency Management 3.1: 96.

Scheper, T., Hitzmann, B., Rinas, U. and Schügerl, K. 1987. Flow cytometry of *Escherichia coli* for process monitoring. Journal of Biotechnology, April 1987, 5(2): 139–148.

Simon Gillespie, Patrick Lipphaus, James Green, Simon Parsons, Paul Weir, Kes Juskowiak, Bruce Jefferson, Peter Jarvis and Andreas Nocker. 1987. Assessing microbiological water quality in drinking water distribution systems with disinfectant residual using flow cytometry. Water Research, Volume 65, 15 November 2014, pp. 224–234.

Stadler, P., Blöschl, G., Vogl, W., Koschelnik, J., Epp, M., Lackner, M., Oismüller, M., Kumpan, M., Nemeth, L., Strauss, P., Sommer, R., Ryzinska-Paier, G., Farnleitner, A.H. and Zessner, M. 2016. Real-time monitoring of beta-d-glucuronidase activity in sediment laden streams: A comparison of prototypes. Water Research 101: 252–261. doi:10.1016/j.watres.2016.05.072.

Starkey, D.E., Han, A., Bao, J.J., Ahn, C.H., Wehmeyer, K.R., Prenger, M.C., Halsall, H.B. and Heineman, W.R. 2001. Fluorogenic assay for beta-glucuronidase using microchip-based capillary electrophoresis. J. Chromatogr. B Biomed. Sci. Appl. 2001 Oct 5; 762(1): 33–41.

Storey, M.V., van der Gaag, B. and Burns, B.P. 2011. Advances in on-line drinking water quality monitoring and early warning systems. Water Res. 45: 741–747.

Total Coliform Rule (TCR)54 FR 27544-27568, 54, 124 June 29 (http://www.epa.gov/ogwdw/disinfection/tcr/pdfs/qrg_tcr_v10.pdf).

van der Linde, K., Lim, B.T., Rondeel, J.M., Antonissen, L.P. and de Jong, G.M. 1999. Improved bacteriological surveillance of haemodialysis fluids: a comparison between Tryptic soy agar and Reasoner's 2A media. Nephrol. Dial Transplant. 14: 2433–2437.

Vogl, W., Koschelnik, J. and Lackner, M. 2013. Keimen frühzeitig auf der Spur. CIT Plus. 12: 15–16.

Vogl, W., Koschelnik, J. and Lackner, M. 2013. Rapid detection of *E. coli* in surface waters for quality and health monitoring using fluorescence-based ColiMinder V. WaterMicro 2013, 7th International Symposium on Health-Related Water Microbiology, September 15–20 (Florianopolis, Brazil).

Wildeboer, D., Amirat, L., Price, R.G. and Abuknesha, R.A. 2010. Rapid detection of *Escherichia coli* in water using a hand-held fluorescence detector. Water Res. 44: 2621–2628.

Wisconsin Beach Health (http://www.wibeaches.us/apex/f?p=181:2).

World Health Organization (WHO). 2006. Guidelines for drinking-water quality. Microbial aspects Vol: 1 (Geneva, Switzerland).

World Health Organization (WHO). 2008. Guidelines for drinking-water quality. Recommendations Vol: 1 (Geneva, Switzerland).

Yuba Shed, Is it safe to swim in the Yuba River? The Yuba River Watershed Information System, http://yubashed.org/pages/it-safe-swim-yuba-river , accessed May 21, 2015.

Zhang, X., Geng, P., Liu, H., Teng, Y., Liu, Y., Wang, Q., Zhang, W., Jin, L. and Jiang, L. 2009. Development of an electrochemical immunoassay for rapid detection of *E. coli* using anodic stripping voltammetry based on Cu@Au nanoparticles as antibody labels. Biosens. Bioelectron. 24: 2155–2159.

8

Real-time Monitoring of Microorganisms in Potable Water Using Online Sensors

Samendra P. Sherchan

1. Introduction

As water quality perturbations related to rapid population growth and industrial activities continue to increase throughout the world, effective water quality monitoring has become critical for water utilities. Due to increasing demand and decreasing supply, the challenge of providing safe drinking water is becoming progressively urgent. Access to high quality water through sustainable treatment and effective water distribution systems (DSs) is essential to contemporary life but unfortunately, many DSs are antiquated and are falling into disrepair, potentially resulting in unacceptable water quality for consumers. On the other hand, recent concerns over not only day to day water quality but also the potential for bioterrorism and infrastructure resiliency have stimulated the development of sensors for real-time monitoring of contaminants in water distribution systems (Meinhardt 2005). According to National Research Council report on direct potable reuse, one of the future research needs is to identify better indicators and surrogates that can be used to monitor process performance and develop online real-time or near real-time analytical monitoring

Department of Global Environmental Health Sciences, Tulane University, New Orleans, LA, 70115.
E-mail: sshercha@tulane.edu

techniques for their measurement. Therefore, there is a significant need for integrated and intelligent sensors to operate in real-time with the ability to monitor intrusion or accidental water contamination (NRC 2012, NRC 2007, Magnuson et al. 2005). For decades, water quality evaluation has been costly, time and labor intensive. Most municipal DSs monitor indicator organisms and are incapable of real-time detection of pathogens that must be performed within minutes. However, in recent years, recognition of these problems has raised concerns by regulatory agencies in attempts to ensure water quality standards and public health security (NRC 2006).

Recent advances in computing and sensor technology have catalyzed progress in real-time monitoring capabilities for water quality. As a result, the ability to monitor water quality has greatly improved. In addition, these advances have greatly enhanced diligent monitoring of pathogens, recognized as a critical need for early warning systems. Currently there is no available technology that will specifically identify a microorganism in real-time; however, there are commercial sensors available that can detect microbial water quality changes in less than five minutes with a sufficient sampling rate as to be considered "real-time". The drinking water community has now explored the use of common water quality sensors for the detection of chemical and biological contamination and to measure water quality parameters such as free chlorine, turbidity, pH, and conductivity. Hence, more water utilities have trended toward recent monitoring programs including continuous data collection using sensors. These data can be accessed through on-site downloading and constant surveillance can be carried out to rapidly detect intentional, natural, or accidental contamination and thus, real-time monitoring provides important early warning information to decision-makers so that they can respond appropriately (USEPA 2010, USEPA 2005). We've reviewed currently available sensors for real-time monitoring of water quality in Table 1. However, there are only three technologies available that can detect microorganisms in real-time.

a) Multi-angle light scattering (MALS) technology
   MALS is a variation of turbidity measurements but instead of one light source, several light sources and angles of refraction are used. With proprietary algorithms, the shape, size, refraction index and internal structure of a particle can be deduced from the light scattering patterns. With this technique, microorganisms can be accurately identified. BioSentry®, which is an inline sensor, uses this technology and it contains a laser beam that strikes individual cells or particles in water, resulting in unique light scattering patterns. Such patterns depend on the size and morphological characteristics of the target particles. Data obtained are

Table 1. Commercial Sensor Characteristics (Modified from Sherchan et al. 2014).

Sensor	Approximate purchase price (US $)	Target	Description
JMAR BioSentry™ Technology: Light Scattering (JMAR, 10905 Technology Place, San Diego, CA 92127)	40,000	Bacteria	The Multi-Angle Light Scattering (MALS) technologies use laser beams to strike individual cells or particles in water within a distribution system, resulting in unique light scattering patterns. Such patterns depend on the morphological characteristics of the target particle. Comparisons of obtained patterns with a computerized database of patterns from known pathogens allows for continuous real-time monitoring and classification of microbial contaminants.
S::CAN spectro::lyser Technology: Light Scattering (s::can Measuring Systems LLC, P.O. Box 36402, Cincinnati, OH 45236)	25,000	TOC, DOC	The S::CAN uses UV-spectroscopy to generate a broadband picture of overall water quality. The assumption is that any new contaminant in the water will be detected as a deviation from the baseline or reference signal. The reference signal is normally generated from historical samples that allow for the system to be trained to a specific type of water. This training is essential for real-time monitoring to reduce the incidence of false alarms.
RMS-W(Real-time Monitoring Systems-Water) (Instant BioScan, LLC, Tucson, AZ, 85745)	40,000	Bacteria	The RMS-W Technology uses intrinsic fluorescence and relies on mie-light scattering plus fluorescence from riboflavin and NADP, giving an indication of the viability of any microbial vegetative cells present in water.
Real Tech Real UVT (1375 Hopkins Street Whitby ON L1N 2C2 Canada)	5,000	UV 254	The Real UVT Online monitor is a continuous UV 254 nm testing monitor. The UV 254 wavelength provides an estimate of organic content in test water. The instrument measures UV transmittance referenced to a test water sample. The UVT online monitor uses two different path lengths to overcome this parameter's typical problems with lamp drift or flow-cell fouling.

Table 1 contd. …

...Table 1 contd.

Sensor	Approximate purchase price (US $)	Target	Description
GE 5310 Online Total Organic Carbon (TOC) unit (GE Analytical Instruments, 6060 Spine Rd., Boulder, CO 80301)	25,000	TOC	The GE TOC unit is a single parameter sensor that measures TOC with selective membrane conductimetric technology. This process separates organic molecules into Cl, CO_2, and SO_4 by an ultraviolet light reactor. These molecules pass into a CO_2 transfer module containing a membrane that only allows CO_2 to pass through. The CO_2 can then be further separated into H^+ and HCO_3. Thereafter the TOC is measured as it accumulates in the conductivity cell.
BACTcontrol, (MicroLAN, 5145 PZ Waalwijk P.O. Box 664 5140 AP The Netherlands)	TBD	Bacteria	It measures the specific enzymatic activities of β-D-glucuronidase as an indicator for the presence of bacterial contamination. The enzyme activity is detected by adding reagents which will fluoresce after being hydrolyzed by the β-D-glucuronidase (GLUC). The grade of activity of the β-D-glucuronidase is determined by an increase in fluorescence per time, which can be translated into amount of bacteria. The reagents are substrate specific for the enzyme to be detected.

compared to patterns with a computerized database of patterns from known pathogens, which are then placed into 4 identifying categories: rods, spores, protozoa, and unknown.

b) Intrinsic fluorescence technology
Instant Bioscan is a microbial detection system that uses Mie Scatter for particle sizing and optical techniques to measure universal "bio-markers" (i.e., NADH and riboflavin) produced by bacteria and fungi. Using the "bio-marker" metabolites and proteins within the microbes, a unique and distinguishable fluorescence can be detected under UV illumination. This intrinsic fluorescence emission is utilized by the sensor as the key marker for detecting microbes.

c) Enzyme-based technology
The BACTcontrol manufactured by microLAN is an on-line automated instrument for the detection of microbiological activity in water. It measures the specific enzymatic activities of β-D-glucuronidase as an indicator for the presence of bacterial contamination. The enzyme activity is detected by adding reagents which will fluoresce after being hydrolyzed by the β-D-glucuronidase (GLUC). The grade of activity of the β-D-glucuronidase is determined by an increase in fluorescence per time, which can be translated into amount of bacteria. The reagents are substrate specific for the enzyme to be detected. The measurements are realized in a short period of time (2–4 hours), in contrast to classical microbiological methods.

To date, there are a limited number of studies that have evaluated the use of commercial water quality sensors for real-time monitoring in DSs (Sherchan et al. 2013, Miles et al. 2011, USEPA 2010, Hall et al. 2007, Allmann et al. 2005, Byer and Carlson 2005, King et al. 2005, Kroll et al. 2005). These studies that evaluated commercial sensors in a DS show that the magnitude of the response depends on the sensor's ability to detect the contaminant and the concentration of the contaminate itself. However, only a few published studies demonstrate how sensors respond to microorganisms; this highlights a challenge for detecting microbial intrusion events. A great majority of commercial sensors detect the chemical composition of drinking water; there are few available sensors for detecting microorganisms in real time (USEPA 2010, USEPA 2006). Based on Sherchan et al. (2013), Miles et al. (2011) and USEPA (2010) evaluation results, sensitivity and threshold levels of these devices need to be further improved before implementing into a SCADA system in a large-scale water quality monitoring program.

However, BioSentry® can also be utilized as a real-time trigger that informs the operator that the water quality is degrading, and that the situation warrants investigation. This is the case when microbial counts in the water increase rapidly. In addition, new software and data management tools need to be upgraded after the design and development of early warning systems for microbial contaminants (Sherchan et al. 2014). A brief comparison of detection systems for microorganisms in water has been described in Table 2. Furthermore, online sensors monitoring the DS have mostly focused on chemical contaminants or basic general water quality parameters. There are very limited online sensors that detect microorganisms in real-time. Byer and Carlson (2005) performed a study evaluating the impact of arsenic, cyanide, and two pesticides on general water quality parameters such as residual chlorine, turbidity, pH, conductivity, and total organic carbon (TOC). Their results showed that cyanide had a measurable influence on all the sensors, whereas arsenic's effect was primarily on sensors measuring electrical conductivity and turbidity. Overall, this study demonstrated that sensors monitoring several general water quality parameters could detect contamination events in a DS. Hall et al. (2007) evaluated six single parameter sensors and three multi-parameter sensors that measured free chlorine, turbidity, pH, specific conductivity, TOC, oxidation reduction potential (ORP), chloride, ammonia, and nitrate. This study assessed the response to contaminants qualitatively using non-chlorinated secondary wastewater effluent. The sensors were challenged with potassium ferricyanide, a pesticide formulation, an herbicide formulation, arsenic trioxide, and nicotine, as model chemical contaminants. *E. coli* K-12 strain in growth media was also tested as a model

Table 2. Comparison of Detection Systems for Microbial Pathogens in Water (Sherchan et al. 2014).

Characteristics of test method	**Online sensors**	**Traditional microbiology**	**Molecular techniques (PCR)**	**ATP Luminescence**	**Immunoassays**
Sample Type	Continuous flow	Grab sample	Grab sample	Grab sample	Grab sample
Assay time	Minutes	Hours to days	Hours	Minutes	Hours
Performance Approach	Full Automation	Manual	Semi-automated	Semi-automated	Semi-automated
Remote Operation	✓	×	×	×	×
Automated Notification	✓	×	×	×	×
Customized Thresholds	✓	×	×	×	×
SCADA compatible	✓	×	×	×	×
Consumable/ Reagent costs	Low Cost	Low Cost	High Cost	Low Cost	High Cost

microbial contaminant. Results showed that no single sensor was able to respond to all the contaminants tested, although the specific conductivity, TOC, free chlorine, chloride, and the ORP sensors did respond to a large number of contaminants.

Sherchan et al. (2013) and Miles et al. (2011) also evaluated commercially sensors including the BioSentry, which is a continuous online microbial sensor that can measure microorganisms in real-time. Based on their studies, the limit of detection for *E. coli* and *Bacillus* spores were 10^3 cfu/mL and 10^2 spores/mL respectively. They also mentioned that there is another problem while monitoring water quality 24/7 hours; it generates an enormous amount of data that must be managed. Furthermore, water utilities need to integrate SCADA system that allows an operator at a master facility to collect data and manage it. The SCADA system can be connected to CANARY (water quality event detection system developed by EPA). Then, the CANARY algorithms can set alarms on online water quality monitors by utilizing "set points" or "trigger levels" which are identified when water quality parameters are outside of an expected range of values. This will allow an operator to make a proper decision before there is an outbreak.

2. Conclusion and Perspectives

Currently, the development of various sensors and online monitoring systems is progressing rapidly. These technologies can have clear and multiple benefits for water utilities, such as lower costs and real-time detection. However, many of these technologies need to be improved further and have not been tested in real world scenarios. With most new technologies, there are still problems with robustness, sensitivity, precision, reproducibility, and reliability. On the other hand, most online sensors are utilized for physicochemical parameters such as TOC, turbidity, pH, and water temperature. Other parameters include free chlorine, fluoride, spectral adsorption, and conductivity. Sensors for microbial contaminants are less frequently utilized by water utilities. Following are major issues related to continuous online sensors:

- False negatives
- False positives
- Limit of detection
- Detection of microbial contaminants via a real-time trigger
- Identification of treatment failures
- Integration of software data management
- Development of SCADA system
- Sensor maintenance & cost evaluation

Keywords: Supervisory control and data acquisition (SCADA) system, large-scale water quality monitoring program, multi-angle light scattering (MALS), Intrinsic fluorescence, enzymatic activity

References

Allmann, T.P. and Carlson, K.H. 2005. Modeling intentional distribution system contamination and detection. Journal AWWA 97(1): 58–71.

Byer, D. and Carlson, K.H. 2005. Real-time detection of intentional chemical contamination in the distribution system. Journal AWWA 97(7): 130–133.

Hall, J.S., Zaffiro, A.D., Marx, R.B., Kefauver, P.C., Krishnan, E.R., Haught, R.C., and Herrmann, J.G. 2007. On-line water quality parameters as indicators of distribution system contamination. Journal AWWA 99(1): 66–77.

Helbling, D.E. and Vanbriesen, J.M. 2008. Continuous monitoring of residual chlorine concentrations in response to controlled microbial intrusions in a laboratory-scale distribution system. Water Res. 42: 3162–3172.

King, K.L. and Kroll, D. 2005. Testing and verification of real-time water quality monitoring sensors in a distribution system against introduced contamination. Proceedings of the AWWA Water Quality Technology Conference, Quebec City, Canada.

Kroll, D. and King, K.L. 2005. Operational validation of an online system for enhancing water security in the distribution system. Proceedings of the AWWA Water Security Congress, Oklahoma City, OK.

Magnuson, M.L. et al. 2005. Responding to water contamination threats. Environ. Sci. Technol. A-Pages 39(7): 153A–159A.

McKenna, S.A., Wilson, M. and Klise, K.A. 2008. Detecting changes in water quality data. Journal AWWA 100(1): 74–85.

Meinhardt, P.L. 2005. Water and bioterrorism; preparing for the potential threat to USA water supplies and public health. Annul. Rev. Public Health 26: 213–37.

Miles, S.L., Sinclair, R.G., Riley, M.R. and Pepper, I.L. 2011. Evaluation of select sensors for real-time monitoring of *Escherichia coli* in water distribution systems. Appl. Environ. Micro. 77(8): 2813–2816.

NRC (National Research Council). 2012. Water Reuse: Potential for Expanding the Nation's Water Supply Through Reuse of Municipal Wastewater; National Academies Press; Washington D.C., USA.

NRC (National Research Council). 2007. Improving the Nation's Water Security: Opportunities for Research; National Academies Press; Washington D.C., USA.

NRC (National Research Council). 2006. Drinking Water Distribution Systems: Assessing and Reducing Risks, National Academies Press: Washington, D.C., USA.

Sherchan, S.P., Gerba, C.P. and Pepper, I.L. 2013. Evaluation of real-time water quality sensors for the detection of intentional bacterial spore contamination of potable water. Journal of Biosensors and Bioelectronics 4: 141–145.

Sherchan, S.P., Kitajima, M., Gerba, C.P. and Pepper, I.L. 2014. Rapid detection technologies for monitoring microorganisms in water. Biosens. J. 3: 109.

USEPA (U.S. Environmental Protection Agency). 2005. Technologies and techniques for early warning systems to monitor and evaluate drinking water quality: A state-of-the art review, EPA/600/R-05/156.
USEPA. 2006. Water Quality Sensor Responses to Potential Chemical Threats in a Pilot-scale Water Distribution System, EPA/600/R-06/068, Washington, DC. (Available through the secure WaterISAC site: www.waterisac.org).
USEPA. 2009. Distribution System Water Quality Monitoring: Sensor Technology Evaluation Methodology and Results, EPA 600/R-09/076, Washington, DC.
USEPA. 2010. Detection of biological suspensions using online detectors in a drinking water distributions system simulator, EPA/600/R-10/005.

Glossary

Maximilian Lackner, Philipp Stadler, Ida Maylen Øverleir, Jacobo Paredes, Imanol Tubía, Sergio Arana, Gregor Tegl, Ana Carolina Cardoso Marques and *Sevcan Aydin*

Note: This glossary was compiled from the chapters in the Handbook, supplemented by important general terms in microbiology. References (http://www.easynotecards.com/print_list/302, ISO 6107 2016, Joan 2013) were used as main resources. More than 500 terms are defined in this glossary. The intention of the glossary is to facilitate readability of this Handbook, and to allow the interested readers to deepen their knowledge on these basic terms.

16S

→ See Mitochondrially encoded 16S RNA

16S ribosomal RNA

16S ribosomal RNA (or 16S rRNA) is the component of the 30S subunit of a prokaryotic ribosome that binds to the Shine-Dalgarno sequence. The genes coding for it are referred to as 16S rRNA gene and are used in reconstructing phylogenies, due to the slow rates of evolution of this region of the gene (David Elliott and Michael Ladomery 2015)

16S rRNA

→ See 16S ribosomal RNA

4MUG

→ See MUG

ABE fermentation

A process that uses bacterial fermentation to produce acetone, n-butanol and ethanol from carbohydrates, such as starch and glucose (Qureshi and Blaschek 2001)

Abiotic factors

Non-living factors that can affect life; examples include soil, nutrients, climate and wind

ABNC

Active but non-culturable (Sachidanandham et al. 2005) → See also VBNC

Acetogenic bacterium

An aerobic, Gram-negative bacterium that is rod-shaped and made of non-sporogenous organisms that produce acetic acid as a metabolic waste product

Acetylene Block Assay

Determines the release of nitrous oxide gas from acetylene-treated soil, which is used to estimate denitrification

Acetylene Reduction Assay

Used to estimate nitrogenase activity by measuring the rate of reduction of ethylene to acetylene

Acidophile

An organism that grows well in an acidic medium

Actinomycete

Gram-positive, non-motile, non-sporing, non-capsulated filaments that break into bacillary and coccoid elements. They resemble fungi and most are free-living, particularly in soil

Actinorhizae

The association which exists between actinomycetes and the roots of plants

Activated sludge

Sludge particles which are produced in raw or settled wastewater, by the growth of organisms in aeration tanks. This happens in the presence of dissolved oxygen. The activated sludge contains living organisms that can feed on incoming wastewater

Activation energy

The amount of energy that is required to bring all molecules, at a given temperature, in one mole of a substance to their reactive state

Active carrier

An infected person who has visible clinical symptoms of a disease and who is capable of transmitting the disease to other individuals

Active site

The location on the surface of the enzyme to which the substrate binds

Adenosine triphosphate

An important energy transfer molecule present in all biological entities

Adjuvant

The material added to an antigen to increase its immunogenicity, e.g., alum

AEC

→ See Anion Exchange Capacity

Aerobic

Term to include organisms that require molecular oxygen (O_2) to survive (aerobic organisms) in an environment that contains molecular oxygen and processes that happen only in the presence of oxygen (aerobic respiration)

Aerobic anoxygenic photosynthesis

Photosynthetic process which takes place under aerobic conditions, but which does not result in the formation of oxygen

Aerotolerant anaerobes

Microbes that can survive in both aerobic and anaerobic conditions because they obtain their energy by fermentation

Aflatoxin

A toxin produced by *Aspergillus flavus* and *Aspergillus parasiticus*, which contaminate groundnut (e.g., peanut) seedlings. This is said to be a cause of hepatic carcinoma (David L. Eaton and John D. Groopman 1993)

Agar

A dried hydrophilic, colloidal substance extracted from red algae species. Agar is used as a solid culture medium for bacteria and other microorganisms

Agarose

Obtained from seaweed and used as a resolving medium in electrophoresis. It consists of non-sulfated linear polymer, which contains D-galactose and 3:6-anhydro-L-galactose alternately

Agglutinates

Visible clumps which are formed as a result of an agglutination reaction

Agglutination reaction

The process of clumping together, in suspension of antigen-bearing cells, microorganisms, or particles in the presence of specific antibodies called agglutinins. This leads to the formation of an insoluble immune complex

Airborne transmission

A route of infection transmission wherein the organism is suspended in or spreads its infection by air

Akinete

A resting non-motile, dormant, thick-walled spore state of cyanobacteria and algae

Alcoholic fermentation

A fermentation process that produces alcohol (ethanol) and carbon dioxide from sugars

Alga (plural: algae)

Phototrophic eukaryotic microorganisms, either unicellular or multicellular. These include phaeophyta: brown algae, spirogyra and red algae

Alkalophile

Organism that has an affinity for alkaline media; thus, growing best in such conditions

Allele

The variant form of a given gene. Different alleles can result in different observable phenotypic traits, such as different pigmentation. However, most genetic variations result in little or no observable variation → See also Allotype

Allochthonous flora

Organisms that are not originally found in soil, but reach there by precipitation, sewage, diseased tissue and other such means. Generally, they do not contribute much ecologically

Allosteric site

A non-active site on the enzyme body, to which a non-substrate compound binds. This may result in conformational changes at the active site

Allotype

Any of various allelic variants of a protein, characterized by antigenic differences → See also Allele

Alpha hemolysis

A partial clearing zone, greenish in color, around a bacterial colony that grows on blood agar

Alpha-proteobacteria

One of the five sub-groups of proteobacteria, each with distinctive 16S rRNA sequences. Mostly contains oligotrophic proteobacteria, many of which have distinctive morphological features → See also 16S rRNA and → Proteobacteria

Alternative complement pathway

A pathway of complement activation, including the C3–C9 components of the classical pathway. It is independent of antibody activity

Alveolar macrophage

A highly active and aggressive phagocytic macrophage that is located on the epithelial lining of the lung alveoli. Alveolar macrophages ingest and destroy inhaled particles and microorganisms

Amensalism (antagonism)

A type of symbiosis, wherein one population is adversely affected, while the other one is unaffected

Ames Test

A test which uses bacteria, often a special strain of salmonella, to test chemicals for mutagenicity and carcinogenicity (Bruce N. Ames et al. 1973)

Amino acid

The basic building block of a protein → See also Protein

Amino acid activation

The first stage in the synthesis of proteins, where the amino acid is attached to transfer RNA

Aminoacyl or acceptor site (A site)

The site on the ribosome that contains an aminoacyl-tRNA at the beginning of the elongation cycle during protein synthesis

Ammonia oxidation

A test to evaluate the ammonia oxidation rate for nitrifiers → See also Nitrifier

Ammonification

Liberation of ammonia by microorganisms acting on organic nitrogenous compounds

Amoeba

A small protozoan that is found as a single cell with a nucleus. It changes shape by extruding its cytoplasm, leading to the formation of pseudopodia, by means of which it absorbs food and moves (Nicholas P. Money 2014)

Amoeboid movement

Movement by means of extrusions of the cytoplasm, leading to formation of foot-like processes called pseudopodia

Amphibolic pathways

Metabolic pathways which function both anabolically and catabolically

Amphitrichous

A cell that has a single flagellum at each end

Amphotericin B

An antibiotic substance that is derived from *Streptomyces nodosus*. It is effective against many species of fungi and certain species of *leishmania*

Anabolism

The processes of metabolism that result in the synthesis of cellular components from precursors of low molecular weight (http://www.chem.qmul.ac.uk/iupac/bioinorg/AB.html#20)

Anaerobic

This term refers to organisms which survive in the absence of oxygen (anaerobic organisms), to the absence of molecular oxygen and to processes that occur in the absence of oxygen, like, e.g., anaerobic respiration

Anaerobic digestion

The process by which biodegradable material is decomposed by microorganisms in the absence of oxygen

Anamorph

A stage of fungal reproduction in which cells are asexually formed by the process of mitosis → See also Mitosis

Anaplerotic reactions

Reactions which support to replenish intermediates in the tricarboxylic acid cycle when their reserves are depleted

Anergy

Decreased responsiveness to antigens, to the extent that there is an inability to react to substances that are expected to be antigenic → See also Antigen

Anion Exchange Capacity (AEC)

Total amount of exchangeable anions that a soil can adsorb. The unit of AEC is centimoles of negative charge per kilogram of soil

Annotation

The process of determining the exact location of specific genes in a genome map

Anoxic

A condition or state that is devoid of oxygen

Anoxygenic photosynthesis

A type of photosynthesis where no oxygen is produced. The process is seen in green sulfur bacteria (GSB), red and green filamentous phototrophs (FAPs, e.g., *Chloroflexi*), purple bacteria, Acidobacteria, and heliobacteria (Donald A. Bryant and Niels-Ulrik Frigaard 2006)

Antagonism

→ See Amensalism

Antagonist

A drug that binds to a hormone, neurotransmitter, or another drug, thereby blocking the action of the other substance

Antheridium

The male gametangium found in Phylum *Oomycota* (kingdom Stramenopila) and Phylum *Ascomyta* (kingdom Fungi) → See also Kingdom and → Phylum

Anthrax

An often fatal and infectious disease, caused by ingestion or inhalation of spores of *Bacillus anthracis*, which are normally found in soil. It can be acquired by humans through contaminated wool or animal products or by inhalation of airborne spores (Gregory J. Martin and Arthur M. Friedlander 2015)

Anthropogenic

Manmade, derived from human activities

Antibiosis

Lysis of an organism brought about by metabolic products of the antagonist. This can be caused by enzymes, lytic agents or other toxic compounds → See also Lysis

Antibiotic

A chemical substance produced by a microorganism with the capacity to inhibit the growth of, or kill other microorganisms

Antibody

A Y-shaped protein (immunoglobulin) made by certain white blood cells which are produced by the body's immune system in response to a foreign substance (antigen). The antibody destroys the antigen

Anticodon triplet

A triplet of nucleotides in transfer RNA which is complementary to the codon in messenger RNA → See also Transfer RNA

Antigen

A foreign substance capable of instigating the immune system into action, inciting a specific immune response and of reacting with the products of that response (stimulation of antibody production)

Antimetabolite

A substance that interferes with a specific metabolic pathway, by inhibiting a key enzyme, due its resemblance with the regular enzyme substrate

Antisense RNA

One of the strands of a double-stranded molecule, which does not directly encode the product, but is complementary to it; thus, inhibiting its activity → See also RNA

Antiseptic

A substance which inhibits the growth and development of micro-organisms, but does not need to kill them

Aplanospore

A spore that is formed during asexual reproduction. It is non-flagellated and non-motile

Apoenzyme

A protein part of an enzyme which is separable from the prosthetic group (the so-called coenzyme)

Apoptosis

A pattern of cell death which is often called 'programmed death' or 'suicide of cells', wherein the cell breaks up into fragments. These fragments are then eliminated by phagocytosis. Apoptosis is a protective mechanism, by which the cell prevents spread of infection to other cells by sacrificing itself → See also Apoptosis

Aporepressor

A product of regulator genes. It combines with the corepressor to form the complete repressor

Arbuscule

Special structure formed in the root cortical cells by arbuscular mycorrhizal fungi. The structure formed resembles a tree

Archaean (plural: archaea)

The domain Archaea are single-celled prokaryotic microorganisms that have no organelles. While they were originally classified as Archaebacterial, a three-domain system is now used to classify them as a separate domain

Artificially acquired passive immunity

A type of temporary immunity which results from the introduction of antibodies produced by another organism or by *in vitro* methods, into the body

Ascocarp

The fruiting body (sporocarp) of an ascomycete phylum fungus. It consists of very tightly interwoven hyphae and may contain millions of asci, each of which typically contains four to eight ascospores → See also Hypha

Ascoma (plural: ascomata)

→ See Ascocarp

Aseptic techniques

Procedures that are performed under strictly sterile conditions. These procedures may be laboratory procedures, such as microbiological cultures (Mike Johnston and Jeff Gricar 2010)

Assimilatory nitrate reduction

Reduction of nitrate to compounds like ammonium, for the synthesis of amino acids and proteins

Associative dinitrogen fixation

An enhanced rate of dinitrogen (N_2) fixation brought about by a close relationship between free-living diazotrophic organisms and a higher plant

Associative symbiosis

Interaction between two dissimilar organisms or biological systems; it is normally mutually beneficial

ATP

→ See Adenosine triphosphate

Autogenous Infection

An infection that occurs due to the microbiota of a patient himself

Autolysins

A lysin which originates in an organism and that is capable of destroying its own cells and tissues → See also Lysin

Autoradiography

Making a radiograph of an object or tissue by recording the radiation emitted by it on a photographic plate. The radiation is emitted by radioactive material within the object or tissue

Autotrophic nitrification

The combined nitrification action of two autotrophic organisms, one converting ammonium to nitrite and the other one oxidizing nitrite to nitrate

Auxotroph

A mutated type of organism that requires specific organic growth factors, in addition to the carbon source present in a minimal medium (Robert M. Hoffman 2011)

Axenic

Pure cultures of micro-organisms, i.e., they are not contaminated by any foreign organisms

Axial filament

Found in *spirochetes*, where it is the organ of motility → See also *Spirochetes*

Bacteremia

Presence of bacteria in the blood

Bacteria

→ See Bacterium

Bacterial Artificial Chromosome

A cloning vector that is derived from *E. coli*, which is used to clone foreign DNA fragments in *E. coli*

Bacterial Photosynthesis

A mode of metabolism which is light-dependent and where carbon dioxide is reduced to glucose. Glucose is used for energy production and biosynthesis. It is an anaerobic reaction (Beatrycze Nowicka and Jerzy Kruk 2016)

Bactericide

A substance that kills bacteria

Bacteriochlorophyll

A light-absorbing pigment that is found in phototrophic bacteria, such as green sulfur and purple sulfur bacteria

Bacteriocin

Substances which are produced by bacteria. They kill other strains of bacteria by inducing a metabolic block

Bacteriology

The study of bacteria, which do not have nuclei separated from the rest of the cell by nuclear membranes. For that reason, they are called prokaryotes (while all the other organisms are eukaryotes) → See also Prokaryotes

Bacteriorhodopsin

A protein involved in light mediated ATP synthesis, which contains retinal. It is one of the main characteristics of archaebacteria → See also Retinal and ATP

Bacteriostatic

An agent that inhibits the growth or multiplication of bacteria, but does not kill them

Bacterium (plural: bacteria)

A prokaryotic, single celled organism. The four basic shapes of bacteria are:

- Cocci – spherical
- Bacilli – rod-shaped
- Spirochaete – spiral-shaped
- Vibrio – comma-shaped

Bacteroid

A genus of bacteroides; They are Gram negative, rod-shaped, anaerobic bacteria which are normal inhabitants of the oral, respiratory, urogenital and intestinal cavities of animals and humans

Baeocytes

Reproductive cells formed by cyanobacteria through multiple fission. They are small and spherical in shape

Balanced growth

Microbial growth where all cellular constituents are synthesized at constant rates, in relation to each other

Barophile

An organism that thrives in conditions of high hydrostatic pressure

Barotolerant

An organism that can tolerate high hydrostatic pressure, although it will grow better under normal pressure

Basal body

A cylindrical structure that attaches the flagella to the cell body at the base of prokaryotic or eukaryotic organisms → See also Flagellum, → Prokaryote and → Eukaryote

Basal medium

Allows the growth of many types of microorganisms which do not require special nutrient supplements

Base composition

The proportion of total bases consisting of guanine plus cytosine or thymine plus adenine base pairs → See also DNA

Basidioma

Fruiting body that produces the basidia → see also Basidium

Basidiospore

The sexual spore of the Basidiomycota, which is formed on the basidium → See also Basidiomycota

Basidiomycota

One of two large phyla that, together with the Ascomycota, constitute the subkingdom Dikarya (often referred to as the 'higher fungi') within the kingdom Fungi

Basidium

A special form of sporophore, characteristic of basidiomycetous fungi, on which the sexual spores are usually borne at the tips of slender projections

Batch culture

A culture of microorganisms which is obtained by inoculating a dish containing a single batch of medium

Batch process

A treatment procedure wherein a tank or reactor (fermenter) is filled, the solution is treated and the tank is emptied. Batch processes are mostly used to cleanse, stabilize, or condition chemical solutions for use in industries (Sharratt 1997)

Benthic zone

The ecological region at the lowest level of a water body, including the sediment surface and some sub-surface layers

Beta hemolysis

A clear zone seen around a bacterial colony growing on blood agar

Bioaccumulation

Intracellular accumulation of chemical substances in living tissue

Bioaugmentation

Addition to the microorganism's environment that can metabolize and grow on specific organic compounds

Bioavailability

The extent to which a drug or other substance becomes available to the target tissue after administration

Biochemical Oxygen Demand (BOD)

Also called biological oxygen demand, BOD is a measure of organic pollution in a wastewater sample. It is the amount of dissolved oxygen consumed in five days by biological processes breaking down organic matter. It is a test that measures the oxygen consumed (in mg/L) over five days at 20°C (DIN EN 1899-1)

Biodegradable

The property by which a substance is capable of being degraded by biological processes, like bacterial or enzymatic action. Standard: ISO 13432

Biodegradation

The process of breakdown of substances by chemical reactions

Biofilm

Ecosystem composed of bacterial cells and an extracellular matrix that mainly consists of polysaccharides to encapsulate and protect the microorganisms therein. A biofilm has a high resistance to antimicrobial agents

Biogas

A gas that is produced from the anaerobic (without oxygen) decomposition of organic matter

Bioinsecticide

A pathogen (bacterium, virus or fungus) used to kill or inhibit the activity of unwanted insect pests

Bioluminescence

The production of light in living organisms by the enzyme luciferase

Biomagnification

Increase in the concentration of a chemical substance, as its position progresses in the food chain

Biomarker

A measureable indicator of a distinct biological state or condition

Bioremediation

The use of microbes to break down toxic or unwanted substances for on-site contaminants removal

Biosensor

Instrument or device capable of measuring a biological signal by transducing biological events into a readable signal (physical or chemical measurable parameters). The physicochemical transducer may be optical, electrochemical, thermometric, piezoelectric, magnetic or micromechanical → See also Transducer

Biosynthesis

Production of cellular constituents from simpler compounds

Bio-Tower

A tower filled with plastic particles, e.g., rings, where air and water are forced up the tower by a counterflow movement. It is an attached culture system

Biotransformation

The chemical alterations of a drug, occurring in the body, due to enzymatic activity

Biotrophic

Close associations seen between two different organisms, that work mutually to benefit of each other

Bioventing

A procedure where the subsurface is aerated to enhance biological activity of naturally occurring microorganisms in the soil

Binary fission

A type of asexual reproduction in which the cell divides into two separate daughter cells, each with identical DNA

BOD

→ See Biochemical Oxygen Demand

Budding

A type of asexual reproduction in which an outgrowth forms from the parent cell. It then usually pinches off to form a separate independent cell

Burst size

The number of phages ejected by a host cell over the course of its lytic life cycle → See also Lysis and Phage

Butanediol fermentation

A kind of fermentation found in *Enterobacteriaceae* family, where 2,3-Butanediol is a major product → See also ABE Fermentation

Calvin cycle

The cycle of enzyme-catalyzed dark reactions of photosynthesis that occurs in the chloroplasts of plants and in many bacteria and that involves the fixation of carbon dioxide and the formation of a 6-carbon sugar

Capacitive (electrical) (C)

Physical property of the matter of maintaining an electrical charge within its volume under an applied electric field

Capillary Electrophoresis (CE)

A family of electrokinetic separation methods performed in submillimeter diameter capillaries and in micro- and nanofluidic channels (Philippe Schmitt-Kopplin 2016)

Capsid

The protein coat surrounding a virus

Capsomere

A protein sub-unit of the capsid of a virus

Carbon cycle

The cycle where CO_2 is taken in and converted to organic compounds by photosynthesis or chemosynthesis, after which it is partially incorporated into sediments and then returned to the atmosphere by respiration or combustion

Carbon fixation

Conversion of CO_2 and other single carbon compounds to organic compounds, such as carbohydrates

Carboxyl group

The -COOH group found attached to the main carbon skeleton in certain compounds, like carboxylic acids and fatty acids

Carboxysomes

Polyhedral cell inclusions which form the key enzyme of the Calvin cycle → See also Calvin Cycle

Catabolism

A process by which complex substances are broken down into simpler compounds, often accompanied by the release of energy →See also Anabolism

Catabolite repression

Transcription-level inhibition of inducible enzymes by glucose, or other easily available carbon sources

CBRN

→ See Chemical, biological, radiological, or nuclear (CBRN) contaminants

CE

→ See Capillary Electrophoresis

Cell

The basic unit of all living things

Cell-mediated immunity

Immunity resulting from destruction of foreign organisms and infected cells by the active action of T-lymphocytes on them. It can be acquired by individuals through the transfer of cells

Cellular slime molds

Slime molds with a vegetative phase containing amoeboid cells that come together to form a pseudoplasmodium

Cellulitis

A diffused inflammation of the soft or connective tissue, in which a thin and watery exudate spreads through tissue spaces, often leading to ulceration and abscess formation

Cephalosporin

A group of broad-spectrum, penicillinase-resistant antibiotics, derived from *Cephalosporium*

CFU

→ See Colony forming units

CFU/100 ml

→ See Colony forming units

Chaperonin

Heat shock proteins which oversee correct folding and assembly of polypeptides

Chemical, Biological, Radiological, or Nuclear (CBRN) contaminants

A CBRN contamination differs from a hazardous material (HAZMAT) incident in both scope (i.e., CBRN can be a mass casualty situation) and intent. CBRN incidents are often assumed to be intentional and malicious

Chemiluminescence

The emission of light (luminescence) as the result of a chemical reaction; also called chemoluminescence

Chemoautotroph

Organisms that obtain their energy from the oxidation of inorganic substances and other carbon compounds

Chemoheterotroph

Organisms that obtain their energy and carbon from the oxidation of organic compounds

Chemolithotroph

Organisms that obtain their energy from oxidation of inorganic compounds, which act as electron donors

Chemoluminescence

→ See Chemiluminescence

Chemoorganotroph

Organisms that obtain their energy and electrons from the oxidation of organic compounds

Chemostat

A continuously used culture device, controlled by limited amounts of nutrients and dilution rates

Chemotaxis

Movement of a motile organism under the influence of a chemical. It may be attracted towards or repulsed by the chemical

Chemotrophs

Organisms that obtain their energy by the oxidation of chemical compounds

Chlamydospore

A thick-walled intercalary or terminal asexual spore that is not shed. It is formed by rounding up of a cell

Chlorophyll

A green photosynthetic pigment usually found in organelles called chloroplasts

Chromosome

A long continuous piece of DNA that carries genetic information

Chronic Carrier

An individual carrying a pathogen over an extended period of time

Chronic wound

A wound that fails to heal within a predictible amount of time and remains in the phase of the wound healing process

Chytrid

A fungus belonging to the genus Chytridomycota. It is spherical in shape and has rhizoids, which are short, thin, filamentous branches ressembling fine roots → See also Genus

Cilium (plural: cilia)

A tiny hair-like structure on the surface of some microorganisms or cells which beats rhythmically to either propel trapped material out of the body, for example in the lungs, or make a free-living microbe move

Cilia→ See Cilium

Ciliate

A protozoan that moves with the help of cilia → See also Protozoan

Clarification

The process of purification of water, where suspended material in the water is removed. It can be done by using sedimentation, filtration or by the use of adsorbing chemicals like alum

Clone

Cells which have descended from a single parent cell. Organisms have identical copies of DNA structure that is obtained by replication

Colonization

Establishment of an entire community of microorganisms at a designated site

Colorless sulfur bacteria

A group of nonphotosynthetic bacteria that oxidize sulfur compounds, thereby deriving their energy

Coliforms

Group of bacteria whose primary habitat is the bowel of humans or warmblooded animals → See also *E. coli*

Colony Forming Units (CFU)

A measure of viable bacterial cells in a sample. In direct microscopic counts (cell counting) all cells, dead and living, are counted, but CFU measures only viable cells. For convenience the results are given as CFU/ml (colony-forming units per milliliter) or CFU/100 ml for liquids, and CFU/g (colony-forming units per gram) for solids. CFU can be calculated using Miles and Misra method (Miles et al. 1938) → See also MFU

Combinatorial biology

The process of transfer of genetic material from one microorganism to another. Mostly used to synthesize products such as antibiotics. It is also used in genetic engineering

Cometabolism

Transformation of a substrate by a microorganism without deriving energy or nutrients from the substrate

Competent

The ability to take up DNA → See also DNA

Complementary DNA

A DNA copy of any RNA molecule, like mRNA or tRNA

Complex viruses

Viruses with capsids that are neither icosahedral nor helical. They have a complicated symmetry

Conditional mutations

Mutations occurring only under certain specific conditions

Conductivity (electrical) (σ)

Physical property of the matter of transmitting an electric charge through. It is the inverse of the resistivity and it depends on the atomic structure of the matter. The unit is S/m

Conidiospore

A thin-walled, asexual spore seen on hyphae which is not contained in sporangium

Conjugants

Mating partners that participate in conjugation, which is a type of sexual reproduction. It is seen in protozoans

Conjugative plasmid

A self transmissible plasmid, or a plasmid that can encode all functions required to bring about its conjugation

Consortium

Two or more members working together, where each organism benefits from the other, thus often performing functions that may not be possible to carry out alone

Constitutive enzyme

Enzymes synthesized in the cell, irrespective of the environmental conditions surrounding the cell

Cosmid

A plasmid vector which can be packed in a phage capsid. It is useful for cloning large fragments of DNA → See also DNA and Phage

Culture of inoculum

Stored culture

Cyanobacterium

A photosynthetic, nitrogen fixing bacterium; Cyanobacteria are also called blue-green algae or **cyanophyta** (Sarma 2012)

Cyanophyta

→ See Cyanobacteria

Cyst

Resting stage of certain bacteria and protozoans, wherein the entire cell is surrounded by a protective layer → See also Bacterium and Protozoan

Cytokine

Non-antibody proteins released by a cell when it comes in contact with specific antigens → See also Antigen

Cytoplasm

The protoplasm of a cell, exclusive of the nucleus

Cytoplasmic membrane

A selectively permeable membrane which is present around the cytoplasm of the cell

Decomposer

The name given to some fungi and soil bacteria that break down dead animals and plants and their waste products into simpler substances, called nutrients

Decomposition

Chemical breakdown of a compound into smaller and simpler compounds by microorganisms

Defined medium

A medium with an exactly known and quantified chemical composition

Degradation

Process by which a compound is transformed into simpler compounds

Denitrification

Reduction of nitrate or nitrite into simpler nitrogenous compounds like molecular nitrogen or nitrogen oxides (ISO 6107 2016)

Derepressible enzyme

Enzyme produced in the absence of a specific inhibitory compound

Desoxyribonucleic Acid (DNA)

The store of genetic information inside living cells and many viruses → See also Gene

Dewatering

Process whereby wet sludge, usually conditioned by a coagulant, has its water content reduced by physical means (ISO 6107 2016)

Diagnostic test

A procedure performed to determine the presence of a disease

Diatom

A major group of algae; Diatoms are unicellular and among the most common types of phytoplankton

Diazotroph

Organism capable of using dinitrogen (N_2) as its sole nitrogen source

Dielectric (impedance)

Characteristic of matter that poorly conducts electricity ($\sigma \ll 1$). Under an electric field the matter reorganizes its structure aligning the internal dipoles accordingly

Dielectrophoresis (DEP)

The effect of particle movement caused by the application of a varying electric field. Only dielectric particles are subjected to this phenomenon

Differential medium

A medium with certain indicators, which helps distinguish between different chemical reactions during growth of organisms on it (vlab.amrita.edu)

Diffused air aeration

A diffused air activated sludge plant takes air, compresses it and discharges it with force, below the surface of water

Digestion

Stabilization, by biological processes, of organic matter in sludge, normally by an anaerobic process (ISO 6107 2016)

Dikaryon

When two nuclei are present in the same hyphal compartment (they may be homokaryon or heterokaryon), it is known as dikaryon

Dilution plate count method

A method of estimating the number of viable microorganisms in a sample → See also VBNC

Dinitrogen fixation

Conversion of molecular dinitrogen into ammonia and organic compounds useful in other biological processes

Direct count

Using direct microscopic examination to determine the number of microorganisms present in a given amount of sample

Disinfectant

An agent to kill microorganisms

DNA

→ See Desoxyribonucleic Acid

DNA fingerprinting

Techniques by which possible differences between different DNA samples can be assessed → See also DNA

Dolipore septum

Specialized cross-wall that separates hypha of fungi belonging to the genus Basidiomycota → See also Genus

Domain

The highest level of biological classification which goes beyond kingdoms. The three domains of biological organisms are Bacteria, Eukarya and Archaea → See also Bacterium, → Eukarya, → Archaea and → Kingdom

Doubling time

The time needed for a certain population to double in number. Also called generation time

Drinking water

Water of a quality suitable for drinking purposes, also called potable water (ISO 6107 2016)

Dynamic

The measured change of a parameter's signal over time

E.c.

→ See *Escherichia coli*

EAB

→ See Electrochemically Active Bacteria (EAB)

ECM

→ See Extracellular Matrix (ECM)

EIA

→ See Enzyme immunoassay

Electrochemically Active Bacteria (EAB)

Microorganisms with the ability to send their electrons, produced during microbial respiration, to the outside of their cells

Electrochemistry

Field of study of the relationship between electric currents and chemical reactions. The electrochemical impedance is the opposition of the matter to allow redox reactions (ionic exchanges) by the effect of the electric current. As it is run at different frequencies, it is possible to distinguish between different electrochemical reactions

Electrochromism

Capability of some materials to change their optical properties when a sufficient electrochemical potential is applied

Electrodialysis

Process used for the deionization of water in which ions are removed, under the influence of an electric field, from one body of water and transferred to another across an ion-exchange membrane (ISO 6107 2016)

ELFA

→ See Enzyme Linked Fluorescent Assay

ELISA

→ See Enzyme-Linked Immunosorbent Assay

Endoenzyme

Enzyme that acts along the internal portion of a polymer

Endonuclease

The endoenzyme responsible for breaking the phosphodiester bonds in a nucleic acid molecule

Endophyte

An organism that may be parasitic or symbiotic with a plant that is grown within

Endospore

A cell which is formed by certain Gram-positive bacteria in unfavorable conditions. An endospore is extremely resistant to heat and other harmful agents → See also Gram Stain

Enhanced rhizosphere degradation

Enhanced activity of microorganisms involved with biodegradation of contaminants near plant roots which is brought about by compounds exuded by the plant roots

Enrichment culture

Technique wherein environmental conditions are altered to aid the growth of a specific organism or group of organisms

Enteric bacteria

These are bacteria present in the intestinal tract of humans and other animals. They may be physiologic or pathologic

Enzymatic

Aided by enzymes

Enzyme

A protein that facilitates a biochemical reaction by speeding up the rate at which it takes place within cells; a biocatalyst

Enzyme Immunoassay (EIA)

An assay that uses an enzyme-bound antibody to detect an antigen. The enzyme catalyzes a color reaction when exposed to substrate → See also Antigen

Enzyme Linked Fluorescent Assay (ELFA)

Similar to EIA (Enzyme immunoassay) except the enzyme catalyzes a fluorescence, not a color reaction → See EIA

Enzyme-Linked Immunosorbent Assay (ELISA)

A test that uses antibodies and color change to identify a substance

Enzyme responsive materials

Materials undergoing any kind of changes in response to distinct enzymes

EPA

The United States Environmental Protection Agency (EPA, USEPA) is an agency of the Federal government of the United States which was created for the purpose of protecting human health and the environment

Episome

An extrachromosomal replicating genetic element found in certain bacteria

Epitope

An antigenic determinant of known structure. It is the region of the antigen to which the variable region of the antibody binds

Equivalent circuit

Theoretical representation of the electric behavior of a system (i.e., a biosensor and the biological sample) by a simple electric circuit composed of ideal components. The aim of these systems is to isolate the electric effects of complex samples

Ericoid mycorrhizae

The type of mycorrhizae found in Ericales plants. These hyphae are capable of penetrating cortical cells

***Escherichia coli* (*E. coli, E.c.*)**

A Gram-negative, facultatively anaerobic, rod-shaped, coliform bacterium of the genus*Escherichia* that is commonly found in the lower intestine of warm-blooded organisms (endotherms). Most *E. coli* strains are harmless; used as FIB (fecal indicator bacterium)

***Escherichia coli* 0157:H7**

A serotype of the bacterial species *Escherichia coli* and is one of the Shiga toxin–producing types of *E. coli*

Estuaries

Water bodies located at river ends. They are subjected to tidal fluctuations

Eubacteria

A genus of bacteria belonging to the family *Propionibacteriaceae*, found as *saprophytes* in soil and water → See also Genus and → *Saprophytes*

Eukaryote

A single-celled or multicellular organism which has a true membrane-bound nucleus and membrane bound organelles → See also Prokaryote

Exoenzyme

An enzyme acting outside the cell that secretes it

Exons

The region of a split DNA that codes for RNA → See also RNA

Extracellular Matrix (ECM)

Complex molecular combination of different polysaccharides, proteins and other molecules generated by cells that house them providing a unique microenvironment

Extremophile

A microbe that positively thrives in environments that would kill other organisms → See also Thermophile and → Halophile

Extracellular

Outside the cell

Exudate

A fluid high in protein and cellular debris which has escaped from blood vessels, usually as a result of inflammation

Facultative organism

An organism which is able to adjust to a particular circumstance or has the ability to take up different roles in a process

FC

→ See Fecal coliforms

Fecal Indicator Bacteria (FIB)

Bacteria which are sensitive to a certain antibiotic and thus leave zones of inhibition around cultures which produce the antibiotic; Relatively easily detected bacteria which indicate potential presence of pathogens (https://mi.water.usgs.gov/h2oqual/BactHOWeb.html)

Feedback initiation

Inhibition by an end product of the biosynthetic pathway involved in its synthesis

Filamentous

In the form of very long rods, mostly seen in bacteria. Seen as branching strands in fungi

Fimbria

Short filamentous structure present on a bacterial cell, involved in adhesion of the bacteria to other surfaces it comes in contact with

Fecal coliforms

Subgroup of coliform bacteria that is capable of growth at temperatures as high as 44°C; Fecal coliforms are facultatively anaerobic, rod-shaped, Gram-negative, non-sporulating bacterium. Coliform bacteria in general originate in the intestines of warm-blooded animals. The term 'thermotolerant coliform' is more correct and is gaining acceptance over 'fecal coliform' (UNEP/WHO 1996)

Fecal contamination

Contamination of microorganisms originating from the bowel of humans or warmblooded animals → See also *E. coli*

Fermentation

The conversion of organic compounds, such as carbohydrates into simpler substances by microbes, usually under anaerobic conditions (with no oxygen present). Energy is produced → See also Fermentor

Fermentor

A reactor for biochemical synthesis; also called fermenter

FIB

→ See Fecal Indicator Bacteria

FISH

→ See Fluorescent *In Situ* Hybridization

Flagellum (plural: flagella)

A long thin appendage present on the surface of some cells, such as bacteria and protoctista, which enables them to move

Flow cytometry

A laser- or impedance-based, biophysical technology employed in cell counting, cell sorting, biomarker detection and protein engineering, by suspending cells in a stream of fluid and passing them by an electronic detection apparatus (Sack et al. 2008)

Fluorescence

Light emitted from a substance upon absorption of electromagnetic radiation

Fluorescent antibody

Uses in a laboratory test, wherein antibodies are tagged with fluorescent dye to detect the presence of microorganisms

Fluorescent *In Situ* Hybridization (FISH)

An in situ hybridization procedure that uses fluorescent probes to detect DNA sequences (O'Connor 2008)

Food poisoning

Any illness caused by eating food contaminated by pathogenic microbes

Food spoilage

Changes in appearance, flavour, odour, and other qualities of the food due to microbial growth which causes it to deteriorate and spoil by decay

Frustule

Siliceous wall and protoplasm seen in diatoms → See also diatom

Fungistasis

Suppression of growth of new fungal cells, due to excessive competition for nutrients, or due to the presence of excessive inhibitory compounds in the soil

Fungus (plural: fungi)

A eukaryotic, non-photosynthetic, spore-forming organism. Fungi range from single celled organisms to very complex multicellular organisms. These heterotrophic organisms live as saprophytes or parasites and the group includes mushrooms, yeast and molds. They have a rigid cell wall

GAC

→ See Granulated Activated Carbon

Gas vacuole

Cellular organelle, found only in prokaryotes, which are gas-filled vesicles

Gene

Basic unit of inheritance located on a chromosome. A gene is a piece of deoxyribonucleic acid (DNA) that contains the instructions for the production of a specific protein

Gene cloning

Isolation of a desired gene from an organism and its replication in large amounts. It is used extensively in DNA research

Gene expression

The process which uses the information that is encoded in a gene to assemble a protein molecule. The gene sequence is read in groups that consist of three bases

Gene probe

A strand of nucleic acid which can be labeled and hybridized to a complementary molecule from a mixture of other nucleic acids. It is helpful in DNA sequencing

Generation time

The time required for a population of microorganisms to double in number; also called doubling time

Genetic code

The information on the DNA, which is required for the synthesis of proteins

Genetic engineering

The controlled manipulation of the genes in an organism with the intent of improving it. This is usually done independently of the natural reproductive process. The result is a so-called genetically modified organism (GMO). To date, most of the effort in genetic engineering has been focused on agriculture (Desmond S.T. Nicholl 2008)

Gene therapy

The therapeutic delivery of nucleic acid polymers into a patient's cells as a drug to a treat disease by DNA changes (B01D8X6SOC 2016)

Genus

A taxonomic rank used in the biological classification of living and fossil organisms in biology. In the hierarchy of biological classification, genus comes above species and below family. The scientific name of a genus may be called the generic name or generic epithet

GLUC

An abbreviation for the enzyme β-glucuronidase (GUS, EC 3.2.1.31) → See GUS

Glycosidase

The enzyme responsible for hydrolizing a glucosidic linkage between two sugar molecules

Gram stain

A differential stain that divides bacteria into two groups, as Gram positive and Gram negative, depending on the ability of the organism to retain crystal violet when decolorized with an organic solvent like ethanol

Granulated Activated Carbon (GAC)

The two main types of activated carbon used in water treatment applications are granular activated carbon (GAC) and powdered activated carbon (PAC).

Growth

An increase in the number of cells, and the size and constituents present in the cells

Growth factor

Organic compound essential for growth and is required in trace amounts; it cannot be synthesized by the organism itself

Growth rate

The rate at which growth occurs

Growth rate constant

Slope of log10 of the number of cells per unit volume plotted against time

Growth yield coefficient

Quantity of carbon formed per unit of substrate carbon consumed

GUS

The enzyme β-glucuronidase (GUS, EC 3.2.1.31). GUS is an acid hydrolase. Chromogenic or fluorogenic enzyme substrates are used to detect the enzymes GUS and GUD → See MUG

GUD

The enzyme β-D-galactosidase (GUD). GUD is an acid hydrolase. Chromogenic or fluorogenic enzyme substrates are used to detect the enzymes GUS and GUD → See MUG

Halophile

An organism which thrives, or at least which can survive in a saline environment → See also Extremophile

Halotolerant

An organism which can survive in a saline environment, but does not require a saline environment for growth

Hapten

A substance not inducing antibody formation, but which is able to combine with a specific antibody

HAZMAT

Hazardous material

Heterofermentation

Any fermentation where there is more than one main end product

Heterokaryon

Hypha that contains at least two genetically dissimilar nuclei

Heterolactic fermentation

A kind of lactic acid fermentation, wherein various sugars are fermented into different products

Heterothallic

Hyphae that are incompatible with each other, thus requiring another compatible hypha to mate with, to form a dikaryon or a diploid

Heterotrophic nitrification

The oxidation of ammonium to nitrite and nitrate by heterotrophic organisms → See also Heterotrophic Organism

High Performance Liquid Chromatography (HPLC)

A technique in analytical chemistry used to separate, identify and quantify each component in a mixture

Holomorph

A fungus that consists of all sexual and asexual stages in its life cycle

Homofermentation

A type of fermentation where there is only one type of end product generated

Homokaryon

A fungal hypha containing nuclei which are genetically identical → See also Hypha

Homolactic fermentation

A type of lactic acid fermentation, in which all sugars involved are converted into lactic acid

Homothallic

Hyphae that are self-compatible, that is, sexual reproduction occurs in the same organism by meiosis and genetic recombination. Fusion of these hyphae leads to the formation of dikaryon or diploid

Host

→ See Host Cell

Host Cell

A cell that is infected by a virus or another type of microorganism

HPLC

→ See High performance liquid chromatography

Hybridization

Natural or artificial construction of a duplex nucleic acid molecule by complementary base pairing between two nucleic acid strands derived from different sources

Hydrocarbon

An organic compound containing carbon and hydrogen only

Hydrogen oxidizing bacterium

These are bacteria that oxidize hydrogen for energy and synthesize carbohydrates, using carbon dioxide as their source of carbon in the absence of other organic compounds

Hydrological event

An event that has direct influence on ground- and surface-waters (e.g., precipitation or snowmelt)

Hyperparasite

Parasite that feeds on another parasite

Hyperthermophile

An organism that thrives in temperatures ranging around 80°C or more
→ See also Extremophile

Hypolimnion

This is the dense, bottom layer of water, that lies below the thermocline, in a thermally stratified lake

Hypha (plural: hyphae)

A very fine thread that is the basic structure of filamentous fungi

IBP

Interference Band-Pass filter

Idiolite secondary

Metabolite produced during idiophase

Idiophase

Production phase following the log phase

Immersion oil

Oil that is sed with an 'oil immersion' microscope objective and condenser for maximum resolution

Immunity

The protection mechanism against infections caused by microorganisms or toxins, that is inherent in the body

Immunoblot

Technique for analyzing or identifying proteins via antigen-antibody specific reactions

Immunofluoresence

Technique to determine the location of an antigen or antibody in a tissue section or smear by fluorescence

Immunogen

A substance with the capacity to bring about an immune response

Immunoglobulin

A protein with antibody activity

IMP

(1) Inosine monophosphate
(2) Inosinic acid

Impedance (electrical) (Z)

Physical characteristic of matter that describes the opposition against the electric current under an alternative electric field (it is a complex parameter composed of resistance and reactance). The unit is [Ω]

IMViC

A set of tests for *E. coli*: I: Indole formation; M: produces acid to turn Methyl Red indicator red; V: Voges-Proskauer reaction (production of acetoin); C: utilization of Citrate as sole carbon source → See also *E. coli*

Inactivation factor

N_0/N in sterilization, where N_0 is the initial count and N is the count after sterilization

Indicator bacteria

→ See Fecal Indicator Bacteria

Indicator parameter

A parameter that when monitored can provide qualitative information about another parameter (e.g., abundance). Indicator parameters have similar behavior in a system like the parameter of interest, but correlation is not as high as for proxy parameters. Indicator parameters are usually monitored with less effort or higher temporal resolution than the parameter of interest

Induced

Refers to enzymes produced in response to the presence of a specific substance (inducer)

Induced enzyme

An enzyme which appears in bacteria in response to the presence of a substrate

Inducible enzyme

An enzyme generated in response to an external factor

Infection

Invasion and multiplication of micro-organisms in body tissues, leading to various diseases and disorders

Inflammation

A reaction of tissue to irritation, injury, or infection. It is a beneficial process as it destroys or contains the pathogen within a small area enabling the healing process to begin

Infrared (IR)

The part of the electromagnetic spectrum with a wavelength from 0.75 μm to approx. 100 μm (near IR from 0.75–2.5 μm, mid IR from 2.5–25 μm and far IR beyond)

Inoculate

To treat a medium with microorganisms for the purpose of creating a favorable growth response

Inoculation port

An entry into a fermentor through which an inoculum is added

Inoculum

Microbial culture used to initiate growth in a new (usually larger) container

Insertion

A type of genetic mutation, wherein single or multiple nucleotides are added to DNA

Insertion sequence

The simplest possible type of transposable elements

Integration

The process by which a DNA molecule becomes incorporated into another genome

Interior coil

Coil of tubing used to heat or cool contents of a fermentor → See also Fermentor

Interspecies hydrogen transfer

The process of hydrogen production and consumption reactions, occurring by the interaction of various microorganisms

Intracellular

Inside the cell

Invert sugar

Mixture of glucose and fructose produced by acid or enzyme hydrolysis of sucrose

In Vitro

Related to a biological process: made to occur in a laboratory vessel or other controlled experimental environment rather than within a living organism or natural setting

In Vivo

Inside the living organism

IR

→ See Infrared

Isoenzyme

When two different enzymes act as catalysts for the same reaction, or set of reactions

Isoelectric point

The pH at which a colloidal particle or a protein molecule has zero charge in relation to the surrounding solution

Isolation

A procedure wherein a pure culture of an organism is obtained from a sample or an environment

ISP

Interference Short-Pass filter

Jaccard's coefficient

An association coefficient of numerical taxonomy, which is the proportion of characters that match, excluding those that both organisms lack → See also Taxonomy

Karst

A landscape and aquifer type, which are characterized by soluble rocks (e.g., limestone or gypsum). Karst systems are characterized by rapid discharge response and high aquifer vulnerability

Kingdom

In biology, kingdom (Latin: regnum, plural: regna) is the second highest taxonomic rank below domain. Kingdoms are divided into smaller groups called phyla

Koch's postulates

Laws given by Robert Koch which prove that an organism is the causative agent of a disease

K-Strategy

Ecological strategy where organisms depend on adapting physiologically to the resources available in their immediate environment

Lab-on-Paper technology

First introduced in 2007, by George Whitesides group, as a method for patterning paper to create well-defined, millimeter-sized channels, comprising hydrophilic paper bound by a hydrophobic polymer, photoresist or wax; also called lab-on-a-chip-technology (Guijt 2017)

Lag phase

The time period when there is no increase in the number of microorganisms, seen after inoculation of fresh growth medium (Yangyang Wang and Robert L. Buchanan 2016) → See also Log Phase

Lamella

Structure elements seen in plants as the layers of protoplasmic membranes in chloroplasts that contain photosynthetic pigments

Lectins

Plant proteins with a high affinity for specific sugar residues

Leghemoglobin

Red colored pigments rich in iron, which are produced in root nodules during symbiotic association between rhizobia and leguminous plants

Light compensation point

The point where the rate of respiration is higher than the rate of photosynthesis, which usually occurs at about 1% of sunlight intensity

Lipopolysaccharide (LPS)

Complex lipid structure containing sugars and fatty acids, which is commonly found in most Gram-negative bacteria → See also Gram Stain

Lithotroph

An organism that uses inorganic substrates such as ammonia or hydrogen to act as electron donors in energy metabolism. They may be chemolithotrophs or photolithotrophs → See also Chemolithotroph and → Photolithotroph

LoD

limit of detection

Log phase

The growth of bacteria (or other microorganisms, as protozoa, microalgae or yeasts) in batch culture can be modeled with four different phases: lag phase (A), log phase or exponential phase (B), stationary phase (C), and death phase (D)

The log phase (sometimes called the logarithmic phase or the exponential phase) is a period characterized by cell doubling (Yangyang Wang and Robert L. Buchanan 2016) → See also Lag Phase

Lophotrichous

An organism that has a polar tuft (flagella) → see also Flagella

LPS

→ See Lipopolysaccharide

Luciferin

A class of small-molecule substrates that are oxidized in the presence of the enzyme luciferase to produce oxyluciferin and energy in the form of light (bioluminescence)

Luxury uptake

Uptake of nutrients in excess of what is required by an organism for its normal growth

Lymphatic system

Lymph nodes linked by a network of small tubes spread throughout the body to transport the lymph fluid

Lysis

The rupture and destruction of a cell which results in the loss of cellular contents, and cell death

Lysogeny

An association where a prokaryote contains a prophage and the virus genome is replicated in sync with the chromosome of the host → See also Prokaryote and Prophage

Lysosome

A cell organelle that contains lytic enzymes → See also Lysis

Macronutrient

A substance required in large amounts for normal growth of an individual

Magnetosome

Small particles of magnetite present in cells that exhibit magnetotaxis

Magnetotactic bacteria

Bacteria that orient themselves according to the earth's magnetic field due to the presence of their magnetosomes

Marker

A DNA sequence that has a confirmed physical location on a given chromosome is known as a marker. Markers can be used to identify a correlation between an inherited disease and the genes that are responsible for it

MALS

→ See multi-angle light scattering (Renliang 2015)

Mass flow (nutrient)

The movement of solutes in relation to the movement of water

Mediator

Molecule that can be oxidized and reduced with successive recycling

Medium

A source where microorganisms are grown

Memory cell

A cell which is produced as part of a normal immune response. These cells remember a specific antigen and are responsible for the rapid immune response, production of antibodies, on exposure to subsequent infections by that particular antigen

Mesophile

An organism that thrives in temperatures ranging from 15–40°C

Messenger RNA (mRNA)

Messenger RNA is a form of RNA in which the genetic information from the DNA is conveyed to the ribosome. Each RNA single-stranded molecule corresponds to a DNA strand

Methanogen

Microorganism that produces methane

Methanogenesis

The production of methane by biological reactions

Methanogenic bacterium

Bacteria which produce methane as a by-product of their metabolism

Methanotroph

An organism capable of oxidizing methane

MFU

→ See Modified Fishman Units

MFU/100 ml

→ See Modified Fishman Units

Microaerophile

Microorganisms which grow well in relatively low oxygen concentration environments

Microbe

→ See Microorganism

Microbial electrolysis cell

Cell where an external potential is applied to force electrons and protons to cross the endothermic barrier to form hydrogen gas. The protons migrate to the cathode and get reduced to form H_2 with electrons travelling from the anode under an applied voltage

Microbial fuel cell

Cell that generates electricity from organic compounds through microbial catabolism

Microbial nanowire

Extracellular protein appendage that mediates long-range electron transfer through biofilms in an energy-efficient manner

Microbial Source Tracking (MST)

Understanding the origin of fecal pollution is paramount in assessing associated health risks as well as the actions necessary to remedy the problem while it still exists; MST can help here (Troy M. Scott et al. 2002)

Microbiology

The branch of biology that deals with the structure, function, uses and modes of existence of microscopic organisms. Its roots go back to 1674 when Antony van Leeuwenhoek made very small but quite powerful magnifying lenses and reported seeing tiny "animalcules". Microbiology generally covers bacteriology, virology and phycology → See also Bacteriology, → Virology and → Phycology

Microcosm

A community or any other unit that is representative of a larger community

Microenvironment

The immediate physical and chemical surroundings of a microorganism

Microfauna

Protozoa, nematodes and anthropods that are smaller than 200 μm

Microflora

This includes bacteria, virus, fungi and algae → See also Bacterium → Virus → Fungus and → Alga

Microfluidic device

Device composed of one or multiple channels and other structures in the microscale range that allows the control of a fluid to perform experimental procedures

Micrometer

1 μm = one millionth of a meter (10^{-6} m)

Micronutrient

Elements that are required for growth in trace amounts, e.g., copper, iron and zinc

Microorganism (micro-organism, microbe)

An organism that is too small to be seen by the naked eye. Also called microbes, these include bacteria, archaea, fungi, protozoa, algae and viruses → See also Microflora → Bacterium → Protozoan → Virus → Fungus and → Alga

Mitochondrially encoded 16S RNA

Mitochondrially encoded 16S RNA (often abbreviated as 16S) is a mitochondrial ribosomal RNA (rRNA). It is the homologue of the prokaryotic 23S and eukaryotic nuclear 28S ribosomal RNAs

Mitosis

A part of the cell cycle when replicated chromosomes are separated into two new nuclei

Mixotroph

Organisms that are capable of assimilating organic compounds as carbon sources, while using inorganic compounds as electron donors

Modified Fishman Units (MFU)

A parameter to indicate the degree of fecal contamination on water samples. MFU takes *E. coli* in VBNC state into account (Vogl et al. 2013) → See also CFU

Mold

A group of saprobic or parasitic fungi causing a cottony growth on organic substances. Mold is multicellular and filamentous → See also Saprobic

Monitoring

Processes and activities (commonly automated) that supervise a system

Monoclonal antibody

Antibody produced from a single clone of cells, which has a uniform structure and specificity

Morphometric characters

Characteristics regarding the depth, dimension, sediment distribution, water currents, etc.

Most Probable Number (MPN)

In general, a method of getting quantitative data on concentrations of discrete items from positive/negative (incidence) data; also known as the method of Poisson zeroes (Oblinger and Koburger 1975). The standard EN ISO 9308-2:2014 describes Water quality—Enumeration of *Escherichia coli* and coliform bacteria - Part 2: Most probable number method (ISO 9308-2:2012)

Motility

The ability of a cell to move from one place to another

Mould

→ See Mold

MPN

→ See Most Probable Number

mRNA

→ See Messenger RNA

MST

→ See Microbial Source Tracking

Mucigel

Gelatinous material found on the surface of roots growing in normal soil

Mucilage

Gelatinous secretions and exudates produced by plant roots and most microorganisms

MUG

4-methylumbelliferyl-β-D-galactoside or 4-methylumbelliferyl-β-D-glucuronide (4MUG, MUG). A fluorogenic substrate to detect enzymes, typically GUS and GUD → See GUS, GUD

MUP

4-methylumbelliferone production, outcome of reaction between the substrate 4-methylumbelliferyl-β-D-galactoside (4MUG, MUG) and the enzyme β-D-galactosidase (GUD) or β-glucuronidase (GUS)

Murein

→ See Peptidoglycan

Mycelium

A branched network of fungal hyphae → See also Hypha

Mycophagous

Organisms which eat fungi

Mycovirus

Viruses which infect fungi

NAD^+

→ See Nicotinamide Adenine Dinucleotide

$NADP^+$

→ Nicotinamide Adenine Dinucleotide Phosphate

NAPL

A non-aqueous phase liquid which may be lighter or denser than water

NASBA

→ See nucleic acid sequence-based amplification

Necrotrophic

A mechanism by which an organism produces lytic enzymes that kill and then break down host cells for its nutrition

Nematode

Eukaryotes that are unsegmented, usually microscopic roundworm

Neutralism

Lack of interaction between two organisms in the same habitat

Niche

Functional role of an organism in a certain habitat

Nicotinamide Adenine Dinucleotide (NAD^+)

An important oxidized coenzyme that is a hydrogen and electron carrier in redox reactions

Nicotinamide Adenine Dinucleotide Phosphate ($NADP^+$)

An important oxidized coenzyme that acts as a hydrogen and electron carrier in various redox reactions

Nitrate reduction (biological)

The process of reduction of nitrate to simpler forms like ammonium by plants and microorganisms

Nitrification

Biological oxidation of ammonium to nitrite and nitrate

Nitrifier

→ See Nitriying Bacteria

Nitrifying bacteria

Chemolithotrophs; Organisms that include species of the genera *Nitrosomonas*, *Nitrosococcus*, *Nitrobacter* and *Nitrococcus*. These bacteria get their energy by the oxidation of inorganic nitrogen compounds. Types include ammonia-oxidizing bacteria (AOB) and nitrite-oxidizing bacteria (NOB)

Nitrogen cycle

The cycle where nitrogen is used by a living organism, then after the organism dies is restored to soil, followed by its final conversion to its original state of oxidation

Nitrogenase

The enzyme required for biological nitrogen fixation

Nodulin

Proteins produced in root hairs or nodules in response to rhizobial infection

Non-growth method

In microbiology, a method where the growth of the organisms is not necessary for possible detection

Northern Blot

Hybridization of single stranded DNA or RNA to RNA fragments

Normal body flora

Microbes that have adapted to living on the body

Nucleic acid

A high molecular weight nucleotide polymer

Nucleic Acid Sequence-Based Amplification (NASBA)

A method in molecular biology which is used to amplify RNA sequences

Nucleoid

The nuclear region of certain organisms like bacteria, which contains chromosomes, but which is not limited by a nuclear membrane

Nucleophilic compound

An electron donor in chemical reactions involving covalent catalysis in which the donated electrons bond with other chemical groups

Nucleus

The control centre of the cell; It contains chromosomes

0157:H7

→ See *Escherichia coli* O157:H7

Oligonucleotide

A short nucleic acid chain, which is obtained from an organism or is synthesized chemically

Oligotroph

A microorganism that has adapted itself to grow in environments that are low in nutrients

On-site methods

Processes and activities that are performed directly at the location of interest

Oospore

Thick-walled spore formed in an oogonium by fungus like organisms like the phylum Oomycota. → See also Oogonium → Fungus and → Phylum

Oogonium (plural: Ooginia)

A small diploid cell which on maturation forms a primordial follicle in a female fetus or the female (haploid or diploid) gametangium of certain thallophytes

Operon

Genes whose expression is controlled by a single operator

Organelle

A membrane enclosed structure, in cells, that has a specialised function

P/A

Presence/absence

Parasitism

Feeding by one organism on the cells of a second, normally larger organism, thus, harming the host

Parasexual cycle

A nuclear cycle wherein genes of haploid nuclei recombine without meiosis

Pasteurization

Process of using heat to kill or reduce the activity of microorganisms in heat-sensitive materials

Pathogen

An organism that is capable of causing an infection, or harming a host cell

Pathogenicity

The ability of a parasite to infect or inflict damage on a host

PCR

→ See Polymerase Chain Reaction

Phage

Also called bacteriophage; A virus that infects and replicates within a bacterium (Dean Watson 2015) → See also Virology

Phagocyte

A white blood cell that can surround, engulf (by phagocytosis) and destroy invading microorganisms including viruses and bacteria. There are two separate groups—macrophages and neutrophils

Photosynthesis

A process that occurs in plants, algae and some bacteria (cyanobacteria) that traps the sun's light energy and uses it to fix carbon dioxide into organic compounds

Phylum (plural: phyla)

In biology, a phylum is a taxonomic rank below kingdom and above class. Traditionally, in botany the term division was used instead of 'phylum' → See also Taxonomy

Phycology

The study of algae, which are microscopic plants capable of photosynthesis. Some algae (e.g., seaweeds) are large, but many others grow as single cells. Many of the blue-green algae (Cyanophyta) resemble bacteria, and some microbiologists consider them as bacteria → See also Cyanophyta

Pellicle

A rigid protein layer just below the cell membrane

Peptidoglycan

Rigid cell wall layer seen in bacteria, also called murein

Peribacteroid membrane

A plant derived membrane which surrounds rhizobia in host cells of legume nodules

Periplasmic space

The area between the cell membrane and cell wall in Gram-negative bacteria → See also Gram Stain

Perithecium

Flask shaped ascocarp open at the tip → See also Ascocarp

Peritrichous flagellation

Multiple flagella present all over the cell surface

Phosphobacterium

Bacteria that are good at dissolving insoluble inorganic phosphate from the soil

Photoautotroph

Self-sufficient organisms that can generate energy from light and carbon dioxide

Photoheterotroph

Organisms able to use light as source of energy and organic materials as carbon source

Photolithotroph

Organism that obtains energy from light and therefore uses inorganic electron donors only to drive its biosynthetic reactions (e.g., carbon dioxide fixation in lithoautotrophs)

Photophosphorylation

Synthesis of high energy phosphate bonds by the use of light as source of energy

Phototaxis

Movement of an organism, or a part of it, towards light

Phycobilin

Water soluble pigment that is seen in cyanobacteria and is the light harvesting pigment for Photosystem II

Pilus

Fimbria like substance present on fertile cells that deals with transfer of DNA during the process of conjugation → See also DNA

Plaque

A localized area of lysis or cell inhibition which is caused due to virus infection → See also Lysis

Plasmogamy

Fusion of two cell contents, inclusive of the cytoplasm and nuclei

Plate Count

Number of colonies formed on a solid culture medium, when uniformly inoculated with a known amount of sample

Point of Care (PoC) test

Medical diagnostic test at the time and place of the patient care

Polar flagellation

The presence of flagella at one or both ends → See also Flagellum

Polymerase Chain Reaction (PCR)

Method of molecular biology to amplify DNA pieces

Potable water

→ See Drinking water

ppbMU

particles per billion of 4-methylumbelliferone

Primary producer

Green plants, algae and cyanobacteria which produce their own food by a process called photosynthesis. They are found at the beginning of the food chain

Primer

A short strand of DNA that is employed in the polymerase chain reaction (PCR) approach

Probe

A single-stranded DNA or RNA sequence that is utilized to detect a sample genome's complementary sequence

Prophage

A bacteriophage (→ See Phage) genome inserted and integrated into the circular bacterial DNA chromosome or existing as an extrachromosomal plasmid; A latent form of a phage, in which the viral genes are present in the bacterium without causing disruption of the bacterial cell (Krupovic et al. 2011)

Proteobacteria

A major phylum of Gram-negative bacteria. They include, e.g., *Escherichia*, *Salmonella*, *Vibrio*, *Helicobacter*, *Yersinia* and many other genera

Protoplast

A cell devoid of a cell wall

Prokaryote

An organism that has a simple cell structure without a membrane-bound nucleus or organelles

Protein

A folded long-chain molecule consisting of amino acids. Each protein has a special function. Proteins are required for the structure, function and regulation of an organism's cell/cells, tissues and organs

Protein

Built from amino acids → See also Enzyme and → Antibody

Prototype

In respect to measuring devices prototype refers to an early product built, usually using novel technology or assay

Protozoan (plural: protozoa)

A eukaryotic, single celled organism that usually lacks chlorophyll

Proxy parameter

A parameter that can be used as a surrogate for another parameter because of their high correlation (commonly $R^2 > 0.95$) and comparable behavior in a system. Proxy parameters are usually monitored with less effort or higher temporal resolution than the parameter of interest

Pseudopodium (plural: pseudopodium)

A temporary extension of the cytoplasm of an amoeboid cell. It is used in both motility and feeding

Pure Culture

A microorganism population of a single strain

QMRA

→ See Quantitative Microbial Risk Assessment

Quantitative Microbial Risk Assessment (QMRA)

The process of estimating the risk from exposure to microorganisms

Quantum yield

Number of defined events occurring per photon absorbed by the system (http://goldbook.iupac.org/Q04991.html)

R2A agar

A culture medium to study bacteria which normally inhabit potable water. These bacteria tend to be slow-growing species and would quickly be suppressed by faster-growing species on a richer culture medium; also called Reasoner's 2A agar (Reasoner et al. 1979)

Radioimmunoassay

An immunological assay that makes use of radioactive antibodies or antigens to detect certain substances → See also Antibody and → Antigen

Rapid method

A method that measures a parameter (and generates accessible data) in high temporal resolution, compared to available standard methods

Reactance (electrical) (X)

The complex component of the impedance. It represents the phase difference between the current and the voltage and is related to the capacity of stored energy, either due to capacitance or inductance. The unit is [Ω]

Reaction center

A photosynthetic complex containing chlorophyll and other compounds

Real-time analysis

Analysis which gives a result in real-time, i.e., immediately after data collection

Reannealing

The process seen on cooling, where two complementary strands of DNA hybridize back into a single strand

Recalcitrant

Being resistant to a microbial attack

Recombination

Process by which genetic elements in two separate genomes are brought together in one unit. This is an important step in gene therapy → See also Gene Therapy

Recycling

A cyclical process by which essential elements are released into the environment where they are then reused

Replication

Conversion of one double-stranded DNA molecule into two identical double-stranded DNA molecules

Resistance (electrical) (R)

The real component of the impedance. It represents the opposition to an alternative electric current and is related to the characteristic of dissipating the electric energy. The unit is [Ω]

Retinal

Also known as retinaldehyde; It is the vitamin A aldehyde

Reverse Transcription

Process of copying information from RNA to DNA → See also RNA and → DNA

Rhizobacteria

Bacteria that are found in roots, where they aggressively colonize

Rhizobia

Bacteria that are capable of living symbiotically in leguminous plant roots, from where they receive energy and commonly fix N_2

Rhizomorph

Mass of fungal hyphae that are organized in long, thick strands with a darkly pigmented outer rind that contains specialized tissues for absorption and water transport → See also Hypha

Ribonucleic acid

→ See RNA

RNA (Ribonucleic acid)

A polymer that is essential in various biological roles in coding, decoding, regulation and expression of genes. RNA and DNA are nucleic acids, and, along with proteins and carbohydrates, form the four major macromolecules essential for all known forms of life. Like DNA, RNA is assembled as a chain of nucleotides, but unlike DNA it is more often found in Nature as a single-strand folded onto itself, rather than a paired double-strand → See also DNA

SAM

→ See Self-Assembled Monolayer

Sanger sequencing

A DNA sequencing approach that was first formally commercialized by Applied Biosystems. This method involves the selective integration of chain-terminating dideoxynucleotides by DNA polymerase during *in vitro* DNA replication

Sanitization

Elimination of pathogenic or harmful organisms, including insect larvae, intestinal parasites and weed seeds

Saprobe

→ See Saprophyte

Saprophyte

Any organism that lives on dead organic matter, as certain fungi and bacteria; also called saprobe

SCADA

→ See supervisory control and data acquisition

Sclerotium

Modified fungal hyphae that form a compact and hard vegetative resting structure with a thick, pigmented outer rind

Secondary metabolite

Product of intermediary metabolism released from a cell, for example, antibiotic

Selective medium

A medium that is biased in allowing only certain types of microorganisms to grow

Self-Assembled Monolayer (SAM)

Organized stack of molecules (forming a monolayer) that are adsorbed (or immobilized) onto a surface (usually a metallic layer like gold) to provide a specific function (commonly biorecognition (Rudra et al. 2017))

Sensor

An object detecting events or changes in the environment. A sensor is capable of converting this data into a measureable output

Serialdilution

Series of stepwise dilutions, normally done in sterile water, which is done to reduce microorganism populations to manageable numbers

Serology

Study of reactions that take place between antigens and antibodies *in vitro*
→ See also Antigen → Antibody and → *In Vitro*

Sheath

Tubular structure that is found either around a chain of cells or around a bundle of filaments

Siderochromes

The compounds which are responsible for iron uptake; they are synthesized by the microorganisms themselves

Siderophore

A metabolite that is formed by some microorganisms; it forms a strong coordination compound with iron

Slime layer

A diffused layer found immediately outside the cell wall in certain bacteria

Slime mold

Microorganisms that are eukaryotic and which lack cell walls → See also Eukaryote

Solarization

A technique to control the growth of pathogens, wherein a plastic sheet is used to cover moistened soil in hot climates, thereby trapping the incoming radiation

Specific activity

The amount of enzyme activity units per mass of protein, expressed as micromoles formed per unit time per milligram of protein

Spectral absorption coefficient

Surrogate parameter for fecal pollution in water capable of giving a real-time and early result

Spermosphere

The area seen around a germinating seed, where there is increased microbiological activity

Spirochetes

Coiled bacteria, of the order *Spirochaetales*, most of which are pathogenic to both humans and animals

Sporangium (plural: sporagia)

A sac containing spores that develop from the fruiting body of a fungus

Spore

A general term for a dormant stage in an organism's life cycle. Spores enable survival of adverse conditions, distribution and reproduction. There are many types which may be produced both asexually and sexually

Spread plate

A technique for performing a plate count of microorganisms → See also Plate Count

Sterilization

The process whereby an object or surface is rendered free of any living microorganisms

Storage polysaccharide

The energy reserves which are stored in a cell when there is excess of carbon available

Strain

Population of cells, all of which arise from a single pure isolate

Substrate

A base on which an organism is grown. They can also be the substances on which compounds and enzymes act; more generally, a reactant consumed in an enzymatic (or catalytic) reaction

Sulfur cycle

The cycle wherein sulfur is taken up by living organisms, then released upon the death of the organism, before being converted to its final state of oxidation

Surrogate parameter

Parameter related to and used instead of another one because it is more easily measurable

Symbiosis

Two dissimilar organisms, living together. Their association may be commensal or mutualistic

Synergism

Association between two organisms that is mutually beneficial

Syntrophy

Interaction between two or more populations that supply each other's nutritional needs

Systemic

Something that involves the entire body and is not localized in the body

Taxonomy

A classification in a hierarchical system

TC

→ See Total Coliforms

TCR

→ See Total Coliform Rule

Teichoic acids

All wall, membrane or capsular polymers containing glycerophosphate or ribitol phosphate residues

Telemorph

One of the stages of sexual reproduction, wherein cells are formed by meiosis and genetic recombination

Temperate virus

A virus that does not cause destruction and lysis of the cells of its host, but instead, its genome may replicate in sync with that of the host

Terminal electron acceptor

The last acceptor of the electron, as it exits the electron transport chain

ThC

→ See Thermotolerant Coliforms

Thermophile

An organism that grows best at elevated temperatures, typically between 45 and 80°C; → See also Extremophile

Thermotolerant Coliforms (ThC)

→ See Fecal Coliforms

Ti plasmid

A conjugative tumor-inducing plasmid that can transfer genes into plants; Seen in the bacterium *Agrobacterium tunefaciens*

Total Coliforms (TC)

A group of bacteria commonly found in the environment, for example, in soil or vegetation, as well as the intestines of mammals, including humans. Total coliform bacteria are not likely to cause illness, but their presence indicates that your water supply may be vulnerable to contamination by more harmful microorganisms. By contrast, *Escherichia coli (E. coli)* is the only member of the total coliform group of bacteria that is found only in the intestines of mammals, including humans

Total Coliform Rule (TCR)

Being a National Primary Drinking Water Regulation (NPDWR) of the USA, this rule was published by EPA in 1989 and became effective in 1990. The rule set both a health goal (Maximum Contaminant Level Goal (MCLG))

and legal limits (Maximum Contaminant Levels (MCLs)) for the presence of total coliforms in drinking water (EPA 2017) → See EPA

Total Viable Count (TVC)

Not a specific micro-organism, but rather a test which estimates total numbers of viable ('living') individual microorganisms present in a set volume of sample. The TVC count may include bacteria, yeasts and mould species. There are different test parameters for different types of samples

Toxin

A foreign substance present in the body, which is mostly generated by microorganisms, that is capable of inflicting damage on the host cell

Transcription

The process by which the RNA of a gene sequence is copied

Transducer (biosensor)

Element capable of translating a non-measurable signal into something measurable (e.g., translating an antigen-antibody binding into a change in conductivity)

Transduction

The process where host genetic information is transferred through an agent, like a virus or a bacteriophage

Transfer RNA (tRNA)

Transfer RNA is a minute form of RNA molecule that facilitates the process by which the messenger RNA is decoded into a protein

Transgenic

Genetically modified plants or organisms, which contain foreign genes, which have been inserted by means of recombinant DNA techniques

Transposable element

A genetic element that can be transposed from one site on a chromosome to another

Transposon

Transposable element which, in addition to transposable genes, carries other genes

Transposon mutagenesis

A mutant phenotype is formed by inactivation of the host gene, which occurs due to the insertion of a transposon

Tricarboxylic acid cycle

A series of metabolic reactions, by which pyruvate is oxidized to carbon dioxide

tRNA

→ See Transfer RNA

Trophic level

Describes the residence of nutrients in various organisms along a food chain, ranging from the primary nutrient assimilating autotrophs to carnivorous animals

Tuft

→ See Flagella

Turbidity

Cloudiness of a fluid caused by suspended solids. It can be obtained by shining light through a liquid and then measuring the scattering

TVC

→ See Total Viable Count

Uronic acid

A class of acidic compounds that contain both carboxylic and aldehydic groups and are oxidation products of sugars. They occur mainly in polysaccharides

Vaccine

A special type of medicine that is given to both people and animals to artificially increase immunity to a particular disease and to prevent an infectious disease from developing

VBNC

→ See Viable but Nonculturable

Vector

An agent that can carry pathogens from one host to another. It can also denote a plasmid or virus used in genetic engineering to insert genes into a cell

Vegetative cell

A growing or actively feeding form of a cell, in contrast to a spore

Vesicles

Spherical structures formed intra-cellularly by certain arbuscular mycorrhizal fungi

Viable But Nonculturable (VBNC)

Living organisms that cannot be cultured on artificial media; also called 'viable but not culturable'

Viable count

Measurement of the concentration of live cells in a microbial population

Vibrio

A genus of Gram-negative bacteria, possessing a curved-rod shape (comma shape)

Virion

The virus particle and the virus nucleic acid surrounded by a protein coat

Virulence

The degree of pathogenicity of a parasite

Viral envelope

A spiky coat that covers the virus's protein coat or capsid

Virology

The study of viruses, which are submicroscopic packages of nucleic acid which can take control of other cells to produce more viruses. Some viruses attack bacteria specifically. Those are called bacteriophages, or short phages (Jane Flint et al. 2015) → See also Phage

Virus

An infectious particle that relies on the cellular machinery of the host cell to grow and replicate

WHO

→ See World Health Organization

Wild Type

Strain of a microorganism that is isolated from Nature; the native and original form of a gene or organism

Winogradsky column

A glass column that allows growth of microorganisms under conditions similar to those found in nutrient-rich water and sediment. This column contains an anerobic lower zone and an aerobic upper zone

World Health Organization (WHO)

A specialized agency of the United Nations that is concerned with international public health. It was established on 7 April 1948 and is headquartered in Geneva, Switzerland

Woronin body

A spherical structure found in fungi belonging to the phylum Ascomycota, which are associated with the simple pore in the septa separating the hyphal compartments → See also Phylum

Wound

A damaged area of the body, usually involving a breaking of the skin

Xenobiotic

A compound that is foreign to the biological systems

Xerophile

An organism that is capable of growing at low water potentials, that is, in very dry habitats

Yeast

A single-celled fungus

Zymogenous flora

Refers to microorganisms that respond rapidly by enzyme production and growth when simple organic substrates become available

References

Beatrycze Nowicka and Jerzy Kruk. 2016. Powered by light: Phototrophy and photosynthesis in prokaryotes and its evolution. Microbiological Research 186-187: 99–118.

Bruce N. Ames, William E. Durston, Edith Yamasaki and Frank D. Lee. 1973. Carcinogens are Mutagens: A Simple Test System Combining Liver Homogenates for Activation and Bacteria for Detection. PNAS 70(8): 2281–5. doi:10.1073/pnas.70.8.2281.

David Elliott. 2015. Michael Ladomery, Molecular Biology of RNA, Oxford University Press, 2nd edition, ISBN: 978-0199671397.

David L. Eaton and John D. Groopman. 1993. The Toxicology of Aflatoxins: Human Health, Veterinary, and Agricultural Significance, Academic Press, ISBN: 978-0122282553.

Dean Watson. 2015. Bacteriophages: Biological Aspects and Advances, Callisto Reference, ISBN: 978-1632390851.

Desmond S.T. Nicholl. 2008. An Introduction to Genetic Engineering, Cambridge University Press, 3rd edition, ISBN: 978-0521615211.

Donald A. Bryant and Niels-Ulrik Frigaard. 2006. Prokaryotic photosynthesis and phototrophy illuminated. Trends in Microbiology 14(11): 488–496. doi:10.1016/j.tim.2006.09.001.

EPA, Revised Total Coliform Rule and Total Coliform Rule https://www.epa.gov/dwreginfo/revised-total-coliform-rule-and-total-coliform-rule, accessed Jan. 21, (2017).

Gregory J. Martin and Arthur M. Friedlander. 2015. *Bacillus anthracis* (Anthrax). *In*: Mandell, Douglas and Bennett's (eds.). Principles and Practice of Infectious Diseases (8th Edition), Volume 2: 2391–2409.e2.

Guijt, R.M. 2017. Lab on a Chip – Future Technology for Characterizing Biotechnology Products, Reference Module in Life Sciences.

http://goldbook.iupac.org/Q04991.html, accessed Jan. 21, (2017).

http://www.easynotecards.com/print_list/302, accessed Jan. 21, (2017).

https://mi.water.usgs.gov/h2oqual/BactHOWeb.html, accessed Jan. 21, (2017).

ISO 6107. Water Quality Vocabulary, Parts 1–9. International Organization for Standardization, Geneva, Switzerland (2016).

IUPAC Recommendations 1997. Glossary of Terms Used in Bioinorganic Chemistry, http://www.chem.qmul.ac.uk/iupac/bioinorg/AB.html#20, accessed Jan. 21, (2017).

Jane Flint, S., Lynn W. Enquist, Vincent R. Racaniello, Glenn F. Rall and Anna-Marie Skalka. 2015. Principles of Virology, Taylor & Francis Ltd., 4th edition, ISBN: 978-1555819514.

Joan L. Slonczewski. 2013. Microbiology, W W Norton & Co. Export, 3rd edition, ISBN: 978-0393923216.

Krupovic, M., Prangishvili, D., Hendrix, R.W. and Bamford, D.H. 2011. Genomics of bacterial and archaeal viruses: dynamics within the prokaryotic virosphere, Microbiol. Mol. Biol. Rev. 75(4): 610–635. doi:10.1128/MMBR.00011-11.

Mike Johnston and Jeff Gricar. 2010. Sterile Products and Aseptic Techniques for the Pharmacy Technician, Pearson, 2nd edition, ISBN: 978-0135109649.

Miles, A.A., Misra, S.S. and Irwin, J.O. 1938. The estimation of the bactericidal power of the blood. The Journal of Hygiene 38(6): 732–49. doi:10.1017/s002217240001158x.

Nicholas P. Money. 2014. The Amoeba in the Room: Lives of the Microbes, Oxford University Press, ISBN: 978-0199941315.

O'Connor, C. 2008. Fluorescence *in situ* hybridization (FISH). Nature Education 1(1): 171.

Oblinger, J.L. and Koburger, J.A. 1975. Understanding and teaching the most probable number technique. J. Milk Food Technol. 38(9): 540–545.

Philippe Schmitt-Kopplin. 2016. Capillary Electrophoresis: Methods and Protocols (Methods in Molecular Biology), Humana Press, 2nd edition, ISBN: 978-1493964017.

Qureshi, N. and Blaschek, H.J. 2001. J. Ind. Microbiol. Biotech. 27: 287. doi:10.1038/sj.jim.7000114.

Reasoner, D.J., Blannon, J.C. and Geldreich, E.E. 1979. Rapid seven-hour fecal coliform test. Appl. Environ. Microbiol. 38(2): 229–36.

Renliang Xu, Light scattering: A review of particle characterization applications, Particuology, Volume 18, February 2015, pp. 11–21.

Robert M. Hoffman. 2011. Tumor-seeking Salmonella amino acid auxotrophs. Current Opinion in Biotechnology 22(6): 917–923.

Rudra, J.S., Kelly, S.H. and Collier, J.H. 2017. Self-Assembling Biomaterials, Reference Module in Materials Science and Materials Engineering.

Sachidanandham, R., Gin, K.Y. and Poh, C.L. 2005. Monitoring of active but non-culturable bacterial cells by flow cytometry. Biotechnol. Bioeng. Jan 5; 89(1): 24–31.

Sack, U., Tárnok, A. and Rothe, G. 2008. Cellular Diagnostics: Basic Principles, Methods and Clinical Applications of Flow Cytometry, Karger, S, ISBN: 978-3805585552.

Sarma, T.A. 2012. Handbook of Cyanobacteria, CRC Press, ISBN: 978-1578088003.

Sharratt, P.N. 1997. Handbook of Batch Process Design, Springer, 1997 edition, ISBN: 978-0751403695.

The Open University, Gene Therapy, Amazon Digital Services LLC, Kindle Edition, ASIN: B01D8X6SOC (2016).

Tim Sandle. 2013. Pharmaceutical Microbiology Glossary, CreateSpace Independent Publishing Platform, 2nd edition, ISBN: 978-1492369219.

Troy M. Scott, Joan B. Rose, Tracie M. Jenkins, Samuel R. Farrah and Jerzy Lukasik. 2002. Microbial source tracking: Current methodology and future directions. Appl. Environ. Microbiol. 68(12): 5796–5803. doi:10.1128/AEM.68.12.5796-5803.2002.

vlab.amrita.edu, Selective and Differential Media for Identifying Microorganisms. Retrieved 21 January, from vlab.amrita.edu/?sub=3&brch=73&sim=720&cnt=1 (2017).

Vogl, W., Koschelnik, J. and Lackner, M. 2013. Rapid detection of *E. coli* in surface waters for quality and health monitoring using fluorescence-based ColiMinder V. WaterMicro 2013, 7th International Symposium on Health-Related Water Microbiology, September 15–20, Florianopolis, Brazil.

Water Quality Monitoring—A Practical Guide to the Design and Implementation of Freshwater Quality Studies and Monitoring Programmes Edited by Jamie Bartram and Richard Ballance Published on behalf of United Nations Environment Programme and the World Health Organization, UNEP/WHO ISBN 0 419 22320 7 (1996).

Yangyang Wang and Robert L. Buchanan. 2016. Develop Mechanistic Models of Transition Periods between Lag/Exponential and Exponential/Stationary Phase, Procedia Food Science 7: 163–167.

Index

R

S

T

V

W

9781498764025

#0001 - 220817 - C17 - 234/156/13 [15] - CB - 9781498764025